"If this is philosophy, it works, because Godfrey-Smith is a rare philosopher who searches the world for clues. Knowledgeable and curious, he examines, he admires. His explorations are good-natured. He is never dogmatic, yet startlingly incisive."
—Carl Safina, *The New York Times Book Review*

"A philosopher of science and an experienced deep-sea diver, Godfrey-Smith has rolled his obsessions into one book, weaving biology and philosophy into a dazzling pattern that looks a lot like the best of pop science. He peppers his latest book with vivid anecdotes from his cephalopod encounters . . . [He] relates dramatic stories of mischief made by captive octopuses . . . His project is no less ambitious than to work out the evolutionary origins of subjective experience . . . The result is an incredibly insightful and enjoyable book."   —Meehan Crist, *Los Angeles Times*

"[*Other Minds* is] a terrific mix of Cousteau-esque encounters with [cephalopods] in the wild[,] . . . wide-ranging scientific discussion, and philosophical analysis. Beautifully written, thought-provoking, and bold, this book is the latest, and most closely argued, salvo in the debate over whether octopuses and other cephalopods are intelligent, sentient beings."
—Olivia Judson, *The Atlantic*

"Peter Godfrey-Smith's *Other Minds* sells us on the sentient cephalopod and the history of our own consciousness, one tentacle at a time."   —Sloane Crosley, *Vanity Fair*

"Fascinating . . . After reading this book, to paraphrase Byron, you will 'love not man the less, but cephalopods more.'"
—Callum Roberts, *The Washington Post*

"A smoothly written and captivating account of the octopus and its brethren . . . [Godfrey-Smith] stresses their dissimilarity to us and other mammals, but he also wants us to appreciate what we have in common . . . Godfrey-Smith mixes the scientific with the personal, giving us lively descriptions of his dives to 'Octopolis,' a site off the east coast of Australia at which octopuses gather."   —Colin McGinn, *The Wall Street Journal*

"[*Other Minds*'] subject is so amazing, it's hard not to be drawn along, just as Godfrey-Smith was when he extended a hand to an octopus and it reached out to return his touch, echoing his interest."   —Irene Wanner, *The Seattle Times*

"Godfrey-Smith skillfully links the details of evolutionary history and biology to broader philosophical debates about the nature and function of consciousness . . . [*Other Minds*] is a valuable contribution to some of the most basic questions about the origins of conscious life."   —Nick Romeo, *Chicago Tribune*

"Delightful . . . Godfrey-Smith explores the issue from many angles, beginning with a succinct and thoughtful discussion of the evolution of animals, and extending to a look at the octopus's remarkable neurological systems . . . Throughout, Godfrey-Smith intertwines his own keen work observing and filming these animals at a remarkable site off of the coast of Australia he calls 'Octopolis.'"   —Adam Gaffney, *The New Republic*

"Such wondrous creatures deserve a remarkable chronicler. They've found one in Godfrey-Smith . . . *Other Minds* is a superb, coruscating book. It's exciting to see bottom-up philosophy— philosophy that starts on the reef and in the sand and then crawls slowly up towards abstraction. That's how all philosophy should be done."   —Charles Foster, *Literary Review*

"Fascinating and often delightful . . . This book ingeniously blends philosophy and science to trace the epic journey from single-celled organisms of 3.8 billion years ago to the awakening and development of cephalopod consciousness."
—Damian Whitworth, *The Times* (London)

"Peter Godfrey-Smith, a philosopher, skilfully combines science, philosophy and his experiences of swimming among these tentacled beasts to illuminate the origin and nature of consciousness."
—*The Economist*

"Godfrey-Smith has set himself a double challenge with this book: (i) putting together what is known about octopi behavior and cognition and (ii) showing why this information challenges our philosophical and scientific conceptions of the mind. The result is most convincing."
—Ophelia Deroy, *Science*

"A concise and elegant guide to evolution, consciousness, and marine biology."
—Gary Drevitch, *Psychology Today*

"Deftly blending philosophy and evolutionary biology . . . Godfrey-Smith couples his philosophical and scientific approach with ample and fascinating anecdotes as well as striking photography from his numerous scuba dives off the Australian coast . . . [*Other Minds*] is both thoroughly enjoyable and informative."
—*Publishers Weekly*

"[*Other Minds* is] an engrossing blend of avidly described underwater adventures . . . and a fluid inquiry into the brain-body connection . . . Godfrey-Smith performs an exceptionally revealing deep dive into the evolutionary progression from sensing to acting to remembering to the coalescence of the inner voice, thus tracking the spectrum between sentience and consciousness."
—Donna Seaman, *Booklist*

"I love this book, its masterful blend of natural history, philosophy, and wonder. *Other Minds* takes us on an extraordinary deep dive, not only beneath the waves for a revelatory and intimate view of the mysterious and highly intelligent octopus, but also through the aeons to look at the nature of the mind and how it came about. It's a captivating story, and Peter Godfrey-Smith brings it alive in vivid, elegant prose. His ardent and humane passion for the octopus is present on every page. A must-read for anyone interested in what it's like to be an octopus or in the evolution of the mind—ours and the very other, but equally sentient, minds of the cephalopods."

—Jennifer Ackerman, author of *The Genius of Birds*

"One of the greatest puzzles of organic life is how and why certain animals became aware of themselves. Peter Godfrey-Smith uses the octopus as a portal to enter nonhuman consciousness, doing so with great sensitivity and firsthand knowledge."

—Frans de Waal, author of *Are We Smart Enough to Know How Smart Animals Are?*

"Peter Godfrey-Smith delivers a revealing exploration of one— no, two!—of evolution's most critical turns, and one remarkable creature's trailblazing, eight-armed foray into a mental life."

—Jonathan Balcombe, author of *What a Fish Knows*

"One of our species' worst qualities is our insistence on an exclusive pathway to consciousness. Fortunately, Peter Godfrey-Smith has given us a roadmap to a whole new territory of thinking. *Other Minds* is a gracious and generous exploration of this different land, one that will make you rethink the entire notion of sentience."

—Paul Greenberg, *New York Times* bestselling author of *Four Fish* and *American Catch*

PETER GODFREY-SMITH

OTHER MINDS

Peter Godfrey-Smith is a distinguished professor of philosophy at the Graduate Center, City University of New York, and a professor of the history and philosophy of science at the University of Sydney. He is the author of several books, including *Theory and Reality: An Introduction to the Philosophy of Science* and *Darwinian Populations and Natural Selection*, which won the 2010 Lakatos Award. His underwater videos of octopuses have been widely featured in the media.

OTHER MINDS

OTHER MINDS

The Octopus, the Sea, and
the Deep Origins of Consciousness

PETER GODFREY-SMITH

FARRAR, STRAUS AND GIROUX   NEW YORK

Farrar, Straus and Giroux
18 West 18th Street, New York 10011

An excerpt from *Other Minds* originally appeared,
in slightly different form, in *Scientific American*.

All photographs were taken by the author, except for the frontispiece
and those on pages 62 (*bottom*), 101, 102, 181, and 183, which are frames
from video taken by unmanned cameras (collaboration by Peter
Godfrey-Smith, David Scheel, Matt Lawrence, and Stefan Linquist).
Drawn figures are by the author with the following exceptions:
page 46, figure by Ainsley Seago; pages 49 and 186, figures by Eliza Jewett.

The Library of Congress has cataloged the hardcover edition as follows:
Names: Godfrey-Smith, Peter.
Title: Other minds : the octopus, the sea, and the deep origins of
consciousness / Peter Godfrey-Smith.
Description: First edition. | New York : Farrar, Straus and Giroux,
2016. | Includes index.
Identifiers: LCCN 2016016696 | ISBN 9780374227760 (cloth) |
ISBN 9780374712808 (e-book)
Subjects: LCSH: Nervous system—Evolution. | Consciousness. |
Cephalopoda—Behavior. | Cephalopoda—Psychology.
Classification: LCC QM451 .G58 2016 | DDC 612.8—dc23
LC record available at https://lccn.loc.gov/2016016696

Paperback ISBN: 978-0-374-53719-7

Designed by Jonathan D. Lippincott

Our books may be purchased in bulk for promotional, educational, or business use.
Please contact your local bookseller or the Macmillan Corporate and Premium
Sales Department at 1-800-221-7945, extension 5442, or by e-mail at
MacmillanSpecialMarkets@macmillan.com.

www.fsgbooks.com
www.twitter.com/fsgbooks • www.facebook.com/fsgbooks

For all those who work to protect the oceans

The demand for continuity has, over large tracts of science, proved itself to possess true prophetic power. We ought therefore ourselves sincerely to try every possible mode of conceiving the dawn of consciousness so that it may *not* appear equivalent to the irruption into the universe of a new nature, non-existent until then.   —William James, *The Principles of Psychology*, 1890

The drama of creation, according to the Hawaiian account, is divided into a series of stages . . . At first the lowly zoophytes and corals come into being, and these are followed by worms and shellfish, each type being declared to conquer and destroy its predecessor, a struggle for existence in which the strongest survive. Parallel with this evolution of animal forms, plant life begins on land and in the sea—at first with the algae, followed by seaweeds and rushes. As type follows type, the accumulating slime of their decay raises the land above the waters, in which, as spectator of all, swims the octopus, the lone survivor from an earlier world.   —Roland Dixon, *Oceanic Mythology*, 1916

CONTENTS

OTHER MINDS

MEETINGS ACROSS THE TREE OF LIFE

Two Meetings and a Departure

On a spring morning in 2009, Matthew Lawrence dropped the anchor of his small boat at a random spot in the middle of a blue ocean bay on the east coast of Australia, and jumped over the side. He swam down on scuba to where the anchor lay, picked it up, and waited. The breeze on the surface nudged the boat, which started to drift, and Matt, holding the anchor, followed.

This bay is well-known for diving, but divers usually visit only a couple of spectacular locations. As the bay is large and typically pretty calm, Matt, a scuba enthusiast who lives nearby, had begun a program of underwater exploration, letting the breeze carry the empty boat around above him until his air ran out and he swam back up the anchor line. On one of these dives, roaming over a flat sandy area scattered with scallops, he came across something unusual. A pile of empty scallop shells—thousands of them—was roughly centered around what looked like a single rock. On the shell bed were about a dozen octopuses, each in a shallow, excavated den. Matt came down and hovered beside them. The octopuses each had a body about the size of a football, or smaller. They sat with their arms tucked away. They were mostly brown-gray, but their colors changed moment by moment. Their eyes were large, and not too dissimilar to human eyes, except for the dark horizontal pupils—like cats' eyes turned on their side.

The octopuses watched Matt, and also watched one another. Some started roaming around. They'd haul themselves out of their dens and move over the shell bed in an ambling shuffle. Sometimes this elicited no response from others, but occasionally a pair would dissolve into a multi-armed wrestle. The octopuses seemed to be neither friends nor enemies, but in a state of complicated coexistence. As if the scene were not sufficiently strange, many baby sharks, each just six inches or so long, lay quietly on the shells as the octopuses roamed around them.

A couple of years before this I was snorkeling in another bay, in Sydney. This site is full of boulders and reefs. I saw something moving under a ledge—something surprisingly large—and went down to look at it. What I found looked like an octopus attached to a turtle. It had a flat body, a prominent head, and eight arms coming straight from the head. The arms were flexible, with suckers—roughly like octopus arms. Its back was fringed with something that looked like a skirt, a few inches wide and moving gently. The animal seemed to be every color at once— red, gray, blue-green. Patterns came and went in a fraction of a second. Amid the patches of color were veins of silver like glowing power lines. The animal hovered a few inches above the sea floor, and then came forward to look at me. As I had suspected from the surface, this creature was *big*—about three feet long. The arms roved and wandered, the colors came and went, and the animal moved forward and back.

This animal was a giant cuttlefish. Cuttlefish are relatives of octopuses, but more closely related to squid. Those three— octopuses, cuttlefish, squid—are all members of a group called the *cephalopods*. The other well-known cephalopods are nautiluses, deep-sea Pacific shellfish which live quite differently from octopuses and their cousins. Octopuses, cuttlefish, and squid have something else in common: their large and complex nervous systems.

I swam down repeatedly, holding my breath, to watch this animal. Soon I was exhausted, but I was also reluctant to stop, as the creature seemed as interested in me as I was in it (in him? in her?). This was my first experience with an aspect of these animals that has never stopped intriguing me: the sense of mutual *engagement* that one can have with them. They watch you closely, usually maintaining some distance, but often not very much. Occasionally, when I've been very close, a giant cuttlefish has reached an arm out, just a few inches, so it touches mine. It's usually one touch, then no more. Octopuses show a stronger tactile interest. If you sit in front of their den and reach out a hand, they'll often send out an arm or two, first to explore you, and then—absurdly—to try to haul you into their lair. Often, no doubt, this is an overambitious attempt to turn you into lunch. But it's been shown that octopuses are also interested in objects that they pretty clearly know they can't eat.

To understand these meetings between people and cephalopods, we have to go back to an event of the opposite kind: a departure, a moving apart. The departure happened quite some time before the meetings—about 600 million years before. Like the meetings, it involved animals in the ocean. No one knows what the animals in question looked like in any detail, but they perhaps had the form of small, flattened worms. They may have been just millimeters long, perhaps a little larger. They might have swum, might have crawled on the sea floor, or both. They might have had simple eyes, or at least light-sensitive patches, on each side. If so, little else may have defined "head" and "tail." They did have nervous systems. These might have comprised nets of nerves spread throughout the body, or they might have included some clustering into a tiny brain. What these animals ate, how they lived and reproduced—all are unknown. But they had one feature of great interest from an evolutionary point of view, a feature visible only in retrospect. These creatures were the last common ancestors of yourself and an octopus, of mammals and cephalopods. They're the "last" common ancestors in the sense of *most recent*, the last in a line.

The history of animals has the shape of a tree. A single "root" gives rise to a series of branchings as we follow the process forward in time. One species splits into two, and each of those species splits again (if it does not die out first). If a species splits, and both sides survive and split repeatedly, the result may be the evolution of two or more clusters of species, each cluster distinct enough from the others to be picked out with a familiar name—*the mammals, the birds.* The big differences between animals alive now—between beetles and elephants, for example—originated in tiny insignificant splits of this sort, many millions of years ago. A branching took place and left two new groups of organisms, one on each side, that were initially similar to each other, but evolved independently from that point on.

You should imagine a tree that has an inverted triangular, or conical, shape from far away, and is very irregular inside—something like this:

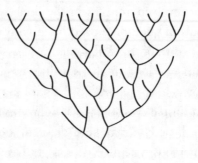

Now imagine sitting on a branch on top of the tree, looking down. You are on the top because you're alive now (not because you are superior), and around you are all the other organisms alive now. Close to you are your living cousins, such as chimpanzees and cats. Further away, as you look horizontally across the top of the tree, you'll see animals that are more distantly related. The total "tree of life" also includes plants and bacteria and protozoa, among others, but let's confine ourselves to the animals. If you now look down the tree, toward the roots, you'll see your ancestors, both recent ones and those more remote. For any pair of animals alive now (you and a bird, you and a fish, a bird and a fish), we can trace two lines of descent down the tree until they meet in a *common* ancestor, an ancestor of both. This common ancestor might be encountered just a short way down the tree, or further down. In the case of humans and chimps we reach a common ancestor very quickly, living about six million years ago. For very different pairs of animals—human and beetle—we have to trace the lines further down.

As you sit in the tree, looking across at your near and distant relatives, consider a particular collection of animals, the ones we usually think of as "smart"—the ones with large brains, who are complex and flexible in their behavior. These will certainly include chimps and dolphins, also dogs and cats, along with humans. All these animals are quite near to you on the tree. They are fairly close cousins, from an evolutionary point of view. If we're doing this exercise properly we should also add birds. One of the most important developments in animal psychology over the last few decades has been the realization of how smart crows and parrots are. Those are not mammals, but they are vertebrates, and hence they are still fairly close to us, though not nearly as close as chimps. Having collected all these birds and mammals, we can ask: What was their most recent common ancestor like, and when did it live? If we look down the tree to where their lines of ancestry all fuse, what do we find living there?

The answer is a lizard-like animal. It lived something like 320 million years ago, a bit before the age of the dinosaurs. This animal had a backbone, was of reasonable size, and was adapted to life on land. It had an architecture similar to our own, with four limbs, a head, and a skeleton. It walked around, used senses similar to ours, and had a well-developed central nervous system.

Now let's look for the common ancestor that connects this first group of animals, which includes ourselves, to an octopus. To find this animal we have to travel much further down the branches. When we find it, about 600 million years before the present, the animal is that flattened worm-like creature I sketched earlier.

This step back in time is nearly twice as long as the step we took to find the common ancestor of mammals and birds. The human-octopus ancestor lived at a time when no organisms had made it onto land and the largest animals around it might have been sponges and jellyfish (along with some oddities I'll discuss in the next chapter).

Assume we've found this animal, and are now watching the departure, the branching, as it happened. In a murky ocean (on the sea floor, or up in the water column) we're watching a lot of these worms live, die, and reproduce. For an unknown reason, some split off from the others, and through an accumulation of happenstance changes they start to live differently. In time, their descendants evolve different bodies. The two sides split again and again, and before long we are looking not at two collections of worms, but at two enormous branches of the evolutionary tree.

One path forward from that underwater split leads to our branch of the tree. It leads to vertebrates, among others, and within the vertebrates, to mammals and eventually humans. The other path leads to a great range of invertebrate species, including crabs and bees and their relatives, many kinds of worms, and also the mollusks, the group that includes clams, oysters, and snails. This branch does not contain all the animals commonly known as "invertebrates," but it does include most of the familiar ones: spiders, centipedes, scallops, moths.

In this branch most of the animals are fairly small, with exceptions, and they also have small nervous systems. Some insects and spiders engage in very complex behavior, especially social behavior, but they still have small nervous systems. That's how things go in this branch—except for the cephalopods. These are a subgroup within the mollusks, so they are related to clams and snails, but they evolved large nervous systems, and the ability to behave in ways very different from other invertebrates. They did this on an entirely separate evolutionary path from ours.

Cephalopods are an island of mental complexity in the sea of invertebrate animals. Because our most recent common ancestor was so simple and lies so far back, cephalopods are an *independent experiment* in the evolution of large brains and complex behavior. If we can make *contact* with cephalopods as sentient beings, it is not because of a shared history, not because of kinship, but because evolution built minds twice over. This is probably the closest we will come to meeting an intelligent alien.

~ Outlines

One of the classic problems of my discipline—philosophy—is the relation between mind and matter. How do sentience, intelligence, and consciousness fit into the physical world? I want to make progress on that problem, vast as it is, in this book. I approach the problem by following an evolutionary road; I want to know how consciousness arose from the raw materials found in living beings. Aeons ago, animals were just one of various unruly clumps of cells that started living together as units in the sea. From there, though, some of them took on a particular lifestyle. They went down a road of mobility and activity, sprouting eyes, antennae, and means to manipulate objects around them. They evolved the creeping of worms, the buzzing of gnats, the global voyages of whales. As part of all this, at some unknown stage, came the evolution of *subjective experience.* For some animals, there's something it *feels like to be* such an animal. There is a self, of some kind, that experiences what goes on.

I am interested in how experience of all kinds evolved, but cephalopods will have special importance in this book. This is firstly because they are such remarkable creatures. If they could talk, they could tell us so much. That is not the only reason they clamber and swim through the book, though. These animals shaped my path through the philosophical problems; following them through the sea, trying to work out what they're doing, became an important part of my route in. In approaching questions about animal minds, it is easy to be influenced too much by our own case. When we imagine the lives and experiences of simpler animals, we often wind up visualizing scaled-down versions of ourselves. Cephalopods bring us into contact with something very different. How does the world look to them? An octopus's eye is similar to ours. It is formed like a camera, with an adjustable lens that focuses an image on a retina. The eyes are similar but the brains behind them are different on almost every scale. If we want to understand *other* minds, the minds of cephalopods are the most other of all.

Philosophy is among the least corporeal of callings. It is, or can be, a purely mental sort of life. It has no equipment that needs managing, no sites or field stations. There's nothing wrong with that—the same is true of mathematics and poetry. But the bodily side of this project has been an important side. I came across the cephalopods by chance, by spending time in the water. I began following them around, and eventually started thinking about their lives. This project has been much affected by their physical presence and unpredictability. It has also been affected by the myriad practicalities of being underwater—the demands of gear and gases and water pressure, the easing of gravity in the green-blue light. The efforts a human must make to cope with these things reflect differences between life on land and in water, and the sea is the original home of the mind, or at least of its first faint forms.

At the start of this book I placed an epigraph from the philosopher and psychologist William James, writing at the end of the nineteenth century. James wanted to understand how consciousness came to inhabit the universe. He had an evolutionary orientation to the issue, in a broad sense that included not just biological evolution but the evolution of the cosmos as a whole. He thought that we need a theory based on continuities and comprehensible transitions; no sudden entrances or jumps.

Like James, I want to understand the relation between mind and matter, and I assume that a story of gradual development is the story that has to be told. At this point, some might say that we already know the outlines of the story: brains evolve, more neurons are added, some animals become smarter than others, and that's it. To say that, though, is to refuse to engage with some of the most puzzling questions. What are the earliest and simplest animals that had subjective experience of some kind? Which animals were the first to *feel* damage, feel it as pain, for example?

Does it feel like something to *be* one of the large-brained cepha-lopods, or are they just biochemical machines for which all is dark inside? There are two sides to the world that have to fit together somehow, but do not seem to fit together in a way that we presently understand. One is the existence of sensations and other mental processes that are felt by an agent; the other is the world of biology, chemistry, and physics.

Those problems won't be entirely resolved in this book, but it's possible to make progress on them by charting the evolution of the senses, bodies, and behavior. Somewhere in that process lies the evolution of the mind. So this is a philosophy book, as well as a book about animals and evolution. That it's a phi-losophy book does not place it in some arcane and inaccessible realm. Doing philosophy is largely a matter of trying to *put things together*, trying to get the pieces of very large puzzles to make some sense. Good philosophy is opportunistic; it uses whatever information and whatever tools look useful. I hope that as the book goes along, it will move in and out of philosophy through seams that you won't much notice.

The book aims, then, to treat the mind and its evolution, and to do so with some breadth and depth. The *breadth* involves think-ing about different sorts of animals. The *depth* is depth in time, as the book embraces the long spans and successive regimes in the history of life.

The anthropologist Roland Dixon attributed to the Hawai-ians the evolutionary tale I used as my second epigraph: "At first the lowly zoophytes and corals come into being, and these are followed by worms and shellfish, each type being declared to conquer and destroy its predecessor . . . ." The story of successive conquests that Dixon outlines is not how the history really went, and the octopus is not the "lone survivor of an earlier world." But the octopus does have a special relation to the history of the mind. It is not a survivor but a second expression of what was present before. The octopus is not Ishmael from *Moby-Dick*, who escaped alone to tell the tale, but a distant relative who came down another line, and who has, consequently, a different tale to tell.

2

A HISTORY OF ANIMALS

Beginnings

The Earth is about 4.5 billion years old, and life itself began perhaps 3.8 billion years ago or so. Animals arrived much later—perhaps a billion years ago, but probably some time after that. For most of the Earth's history, then, there was life, but no animals. What we had, over vast stretches of time, was a world of single-celled organisms in the sea. Much of life today goes on in exactly that form.

When picturing this long era before animals, one might start by visualizing single-celled organisms as solitary beings: countless tiny islands, doing nothing more than floating about, taking in food (somehow), and dividing into two. But single-celled life is, and probably was, far more entangled than that; many of these organisms live in association with others, sometimes in mere truce and coexistence, sometimes in genuine collaboration. Some of the early collaborations were probably so tight that they were really a departure from a "single-celled" mode of life, but they were not organized in anything like the way that our animal bodies are organized.

When picturing this world, we might also presume that because there are no animals, there's no behavior, and no sensing of the world outside. Again, not so. Single-celled organisms can sense and react. Much of what they do counts as *behavior* only in a very broad sense, but they can control how they move and what chemicals they make, in response to what they detect going on around them. In order for any organism to do this, one part of it must be *receptive*, able to see or smell or hear, and another part must be *active*, able to make something useful happen. The organism must also establish a connection of some sort, an arc, between these two parts.

One of the best-studied systems of this kind is seen in the familiar *E. coli* bacteria, which live in vast numbers inside and around us. *E. coli* has a sense of taste, or smell; it can detect welcome and unwelcome chemicals around it, and it can react by moving toward concentrations of some chemicals and away from others. The exterior of each *E. coli* cell has an array of sensors—collections of molecules bridging the cell's outer membrane. That's the "input" part of the system. The "output" part is composed of *flagella*, the long filaments with which the cell swims. An *E. coli* bacterium has two main motions: it can *run* or *tumble*. When it runs, it moves in a straight line, and when it tumbles, as you might expect, it randomly changes direction. A cell continually switches between these two activities, but if it detects an increasing concentration of food, its tumbling is reduced.

A bacterium is so small that its sensors alone can give it no indication of the direction that a good or bad chemical is coming from. To overcome this problem, the bacterium uses time to help it deal with space. The cell is not interested in how much of a chemical is present at any given moment, but rather in whether that concentration is increasing or decreasing. After all, if the cell swam in a straight line simply because the concentration of a desirable chemical was high, it might travel away from chemical nirvana, not toward it, depending on the direction it's pointing. The bacterium solves this problem in an ingenious manner: as it senses its world, one mechanism registers what conditions are like right now, and another records how things were a few moments ago. The bacterium will swim in a straight line as long as the chemicals it senses seem *better* now than those it sensed a moment ago. If not, it's preferable to change course.

Bacteria are one among several kinds of single-celled life, and they are simpler in many ways than the cells that eventually came together to make animals. Those cells, *eukaryotes*, are larger and have an elaborate internal structure. Arising perhaps 1.5 billion years ago, they are the descendants of a process in which one small bacterium-like cell swallowed another. Single-celled eukaryotes, in many cases, have more complicated capacities to taste and swim, and they also edge close to a particularly important sense: vision.

Light, for living things, has a dual role. For many it is an intrinsically important resource, a source of energy. It can also be a source of information, an indicator of other things. This second use, so familiar to us, is not easily achieved by a tiny organism. Much of the use of light by single-celled organisms is for solar power; like plants, they sunbathe. Various bacteria can sense light and respond to its presence. Organisms so small have a difficult time determining the direction light is coming from, let alone focusing an image, but a range of single-celled eukaryotes, and perhaps a few remarkable bacteria, do have the beginnings of *seeing*. The eukaryotes have "eyespots," patches that are sensitive to light, connected to something that shades or focuses the incoming light, making it more informative. Some eukaryotes seek light, some avoid it, and some switch between the two; they follow light when they want to take in energy, and avoid it when their energy supplies are full. Others seek out light when it is not too strong and avoid it when the intensity becomes dangerous. In all these cases, there is a control system connecting the eyespot with a mechanism that enables the cell to swim.

Much of the sensing done by these tiny organisms is aimed at finding food and avoiding toxins. Even in the earliest work on *E. coli*, though, it seemed that something else was going on. They were also attracted to chemicals they could not eat. Biologists who work on these organisms are more and more inclined to see the senses of bacteria as attuned to the presence and activities of other cells around them, not just to washes of edible and inedible chemicals. The receptors on the surfaces of bacterial cells are sensitive to many things, and these include chemicals that bacteria themselves tend to excrete for various reasons— sometimes just as overflow of metabolic processes. This may not sound like much, but it opens an important door. Once the same chemicals are being sensed and produced, there is the possibility of coordination between cells. We have reached the birth of social behavior.

An example is *quorum sensing*. If a chemical is both produced and sensed by a particular kind of bacterium, it can be used by those bacteria to assess how many individuals of the same kind are around. By doing this, they can work out whether enough bacteria are nearby for it to be worthwhile to produce a chemical that does its job only if many cells make it at once.

An early case of quorum sensing to be uncovered involves— appropriately for this book—the sea and a cephalopod. Bacteria living inside a Hawaiian squid produce light by a chemical reaction, but only if enough other bacteria are around to join in. The bacteria control their illumination by detecting the local concentration of an "inducer" molecule, which is made by the bacteria and gives each individual a sense of how many potential light producers are around. As well as lighting up, the bacteria follow the rule that the more of this chemical you *sense*, the more you *make*.

When enough light is being produced, the squid who house the bacteria gain the benefit of camouflage. This is because they hunt at night, when moonlight would normally cast their body's shadow down to predators below. Their internal lights cancel the shadow. Meanwhile, the bacteria seem to benefit from the hospitable living quarters provided by the squid.

This aquatic setting is the right one to have in mind when thinking about these early stages in life's history—though in the evolutionary story we are at a point long before there were any squid. The chemistry of life is an aquatic chemistry. We can get by on land only by carrying a huge amount of salt water around with us. And many of the evolutionary moves made at these early stages—those giving birth to sensing, behavior, and coordination—would have depended on the sea's free movement of chemicals.

So far, all the cells we've met are sensitive to external conditions. Some also have a special sensitivity to *other organisms*, including organisms of the same kind. Within that category, some cells show a sensitivity to chemicals that other organisms *make to be perceived*, as opposed to chemicals made as mere byproducts. That last category—chemicals that are made because they'll be perceived and responded to by others—brings us to the threshold of signaling and communication.

We're arriving at two thresholds, though, not one. In a world of single-celled aquatic life, we've seen how individuals can sense their surroundings and signal to others. But we're about to look at the transition from single-celled life to many-celled life. Once that transition is under way, the signaling and sensing that connected one organism to another become the basis of new interactions which take place *within* the new forms of life now emerging. Sensing and signaling between organisms gives rise to sensing and signaling within an organism. A cell's means for sensing the external environment become a means to sense what other cells within the same organism are up to, and what they might be saying. A cell's "environment" is largely made up of other cells, and the viability of the new, larger organism will depend on coordination between these parts.

~ *Living Together*

Animals are multicellular; we contain many cells that act in concert. The evolution of animals began when some cells submerged their individuality, becoming parts of large joint ventures. The transition to a multicellular form of life occurred many times, leading once to animals, once to plants, on other occasions to fungi, various seaweeds, and less conspicuous organisms. Most likely, the origin of animals did not stem from a meeting between lone cells who drifted together. Rather, animals arose from a cell whose daughters did not separate properly during cell division. Usually, when a single-celled organism divides into two, the daughters go their separate ways, but not always. Imagine a ball of cells that forms when one cell divides and the results stay together—and the process repeats several times. The cells in the clump probably ate bacteria as they hovered together in the sea.

The next stages in the history are unclear; a couple of rival scenarios are on the table, based on different kinds of evidence. In one scenario, perhaps the majority view, some of these balls of cells forsook their suspended life and settled on the sea floor. There they began feeding by filtering water through channels in their bodies; the result was the evolution of the sponge.

A sponge? It seems that one could hardly pick a more implausible ancestor: sponges, after all, do not move. They look like an immediate dead end. However, only the adult sponge is stationary. The babies, or larvae, are another matter. They are often swimmers, who search for a place to settle and become an adult sponge. Sponge larvae have no brains, but they have sensors on their bodies sniffing their world. Perhaps some of these larvae opted to *keep* swimming, rather than settle down. They remained mobile, became sexually mature while suspended in the water, and began a new kind of life. They became the mothers of all the other animals, leaving their relatives fixed to the sea floor.

The scenario I just described is motivated by the view that sponges are the living animals most distantly related to us. *Distant* does not mean *old*; present-day sponges are the products of as much evolution as we are. But for various reasons, if sponges did branch off very early, they are thought to offer clues to what the earliest animals were like. Recent work, however, suggests that sponges might not, after all, be the animals most distantly related to us; instead, this title may belong to the *comb jellies*.

A comb jelly, or *ctenophore*, looks like a very delicate jellyfish. It's an almost transparent globe, with colorful bands of hair-like strands running down its body. Comb jellies have often been seen as cousins of jellyfish, but the observable similarities might be misleading; they might have split off from the line leading to other animals even before sponges did. If this is true, it does not mean that our ancestor looked like a present-day comb jelly. But the comb jelly scenario does motivate a different picture of the early evolutionary stages. Again we start with a clump of cells, but then imagine that this clump folds into a filmy globe-like form, and swims in a simple rhythm as it lives suspended in the water column. The evolution of animals proceeds from there—from a hovering ghost-like mother, rather than a wriggling sponge larva who refused to settle down.

When multicellular organisms arise, the cells that were once organisms in their own right begin to work as parts of larger units. If the new organism is to be any more than a clump of cells glued together, it requires coordination. Earlier I described the forms of sensing and acting seen in single-celled life. In multi-cellular organisms, these sensory and behavioral systems be-come more complicated. Further, the very *existence* of these new entities—animal bodies—depends on those capacities for sens-ing and action. Sensing and signaling between organisms gives rise to sensing and signaling within them. The "behavioral" capacities of cells that once lived as whole organisms become the basis for coordination within the new multicellular organism.

Animals give that coordination several roles. One role is seen also in other multicellular organisms, such as plants: signaling be-tween cells is used to *build* the organism, to bring it into being. Another role exists on a faster time scale, and is especially characteristic of animal life. In all but a few animals, the chemical interactions between some cells become the basis for a *nervous sys-tem*, small or large. And in some of these animals, a mass of such cells concentrated together, sparking in a chemo-electrical storm of repurposed signaling, become a brain.

~ *Neurons and Nervous Systems*

A nervous system is made of many parts, but the most significant are the unusually shaped cells called *neurons*. Their long strands and elaborate branchings form a maze through our heads and bodies.

The activity of neurons depends on two things. One is their electrical excitability, seen especially in the *action potential*, an electrical spasm that moves along a cell in a chain reaction. The other is chemical sensing and signaling. A neuron will release a tiny spray of chemicals into the gap or "cleft" between it and another neuron. These chemicals, when they are detected at the other side, can help trigger (or in some cases suppress) an action potential in that adjoining cell. This chemical influence is the residue of ancient signaling between organisms, pressed inward. The action potential, too, existed in cells before animals evolved, and exists today outside them. The first one ever measured, in fact, was in a plant, the Venus flytrap, at the instigation of Charles Darwin in the nineteenth century. Even some single-celled organisms have action potentials.

What nervous systems make possible is not cell-to-cell signaling itself—that is common—but particular kinds of signaling. Nervous systems are *fast*, first of all. Except in a few cases like the Venus flytrap, plants act on a slower time scale. Second, the neuron's long, tenuous projections enable one cell to reach some distance through the brain or body and affect just a few distant cells; influence is *targeted*. Evolution has transformed cell-to-cell signaling from an activity in which cells simply broadcast their signals to whoever is close enough and listening into something different: an organized network. In a nervous system like our own, the result is a continual electrical clamor, a symphony of tiny cellular fits, mediated by sprays of chemicals across the gaps where one cell reaches out to another.

This internal tumult is also *expensive*. Neurons cost a great deal of energy to run and maintain. Creating their electrical spasms is like the continual charging and discharging of a battery, hundreds of times each second. In an animal like us, a large proportion of the energy taken in as food, nearly a quarter in our case, is spent just keeping the brain running. Any nervous system is a very costly machine. Soon I'll turn to the history of this machine, when it might have evolved and how. First, I'll spend some time on a general question about *why*.

Why is it worth having such a brain, or any nervous system? What are they for? As I see it, two pictures guide people's thinking about the matter. These pictures are visible in scientific work and they permeate philosophy, too; their roots run deep. According to the first view, the original and fundamental function of the nervous system is to link *perception* with *action*. Brains are for the guidance of action, and the only way to "guide" action in a useful way is to link what is done to what is seen (and touched, and tasted). The senses track what's going on in the environment, and nervous systems use this information to work out what to do. I'll call this the *sensory-motor* view of nervous systems and their function.*

Between the senses on one side and the "effector" mechanisms on the other, there must be something that bridges the gap, something that uses the information the senses have gained. Even bacteria have this layout, as the case of *E. coli* showed us. Animals have more complex senses, engage in more complex actions, and possess more complex machinery linking their senses and their actions. According to the sensory-motor view, though, the go-between role has always been central to nervous systems— central at the beginning, central now, and at all stages on the way.

This first view is so intuitive that it might seem there's no room for an alternative. But there is another picture, easier to lose sight of than the first. Modifying your actions in response to events going on outside you has to be done, yes, but something else has to happen, too, and in some circumstances it is more basic and more difficult to achieve. This is *creating actions themselves*. How is it that we are able to act in the first place?

Just above, I said: you sense what's going on and do something in response. But *doing* something, if you are made of many

---

*If you've seen the word "sensorimotor" instead, please treat this as the same.

cells, is not a trivial matter, not something that can simply be assumed. It takes a great deal of coordination between your parts. This is not a big deal if you are a bacterium, but if you're a larger organism, things are different. Then you face the task of generating a coherent whole-organism action from the many tiny outputs—the tiny contractions, contortions, and twitches—of your parts. A multitude of *micro*-actions must be shaped into a *macro*-action.

This is familiar to us in social situations as the problem of teamwork. The players on a football team must combine their actions into a whole, and at least in some kinds of football, this would be a substantial task even if the other team always never varied its moves. An orchestra must solve the same problem. The problem that teams and orchestras face is confronted by some individual organisms, too. This issue is largely peculiar to animals; it's a problem for multicellular organisms, not single-celled ones, and only a problem for those multicellular organisms whose lifestyle involves complex actions. It's not much of a problem for bacteria, and not a big problem for seaweed.

Above I treated interactions between neurons as a kind of signaling. Though the analogy is not complete, it is helpful again here as a way of understanding these two visions of the role of early nervous systems. Recall the story of the ride of Paul Revere at the start of the American Revolution in 1775, as told (with considerable poetic license) by Henry Wadsworth Longfellow. The sexton of the Old North Church in Boston was able to observe the movements of the British Army, and he used a lantern code to send a message to Paul Revere ("one if by land; two if by sea"). The sexton was like a sensor, Revere like a muscle, and the sexton's lantern acted like a nervous connection.

The story of Revere is often used to get people to think about communication in an exact way. And so it does. But it also nudges us toward thinking about a particular kind of communication, which solves a particular kind of problem. Consider a different, though still familiar, situation. Suppose you are in a boat with several rowers, each with one oar. The rowers together can propel the boat forward, but even if they are vigorous, their individual actions will not get the boat to go anywhere unless they coordinate what they're doing. It doesn't matter exactly when they pull their oar, as long as they pull at the same time. One way to deal with this situation is to have someone call the "stroke."

Communication in everyday life serves both roles: there is a sexton-and-Revere or sensory-motor role, based on a division between those who see and those who act, and there is a purely coordinative role, as seen in the rowers. Both of these roles can be played at the same time and there's no conflict between them. Getting a boat to move requires the coordination of micro-actions, but someone also needs to watch where the boat is going. The person calling the stroke, the coxswain or "cox," usually acts as the crew's eyes *and* as a coordinator of micro-actions. The same combination can be seen in a nervous system.

Though there's no essential clash between these roles, the distinction itself is important. Through much of the twentieth century, a sensory-motor view of the evolution of nervous systems was simply assumed, and it took some time for the second view, the one based on internal coordination, to become clear. Chris Pantin, an English biologist, developed the second view in the 1950s and it has been revived recently by Fred Keijzer, a philosopher. They rightly point out that it's easy to fall into the habit of thinking of each "action" as a single unit, in which case the only problem left to solve is coordinating these acts with the senses, working out when to do X rather than Y. As organisms get bigger and can do more, that picture becomes more and more inaccurate. It ignores the problem of how an organism is able to do X or Y in the first place. Pressing an alternative to the sensory-motor theory was a good thing. I'll call this the *action-shaping* view of the role played by early nervous systems.

Returning to the history, what did the first animals with nervous systems look like? How should we picture their lives? We don't yet know. Much of the research in this area has been focused on the *cnidarians* (pronounced "nye-dair-ians"), a group of animals that includes jellyfish, anemones, and corals. They are very distantly related to us, but not as distantly as sponges, and they do have nervous systems. Though the early branchings in the tree of animals remain murky, it is common to think that the animal with the first nervous system might have been jellyfish-*like*—something soft, with no shell or skeleton, probably hovering in the water. Picture a filmy lightbulb in which the rhythms of nervous activity first began.

This might have occurred something like 700 million years ago. That date is based entirely on genetic evidence; there are no fossils of animals this old. From looking at rocks of this age, you'd think that all was still and silent. But DNA evidence strongly suggests that many of the crucial branching points in the history of animals must have occurred around that time, and that means that animals were doing *something* back then. The uncertainty about these crucial stages is frustrating for someone who wants to understand the evolution of brains and minds. As we get a little closer to the present, the picture starts to become clearer.

~ *The Garden*

In 1946, an Australian geologist, Reginald Sprigg, was exploring some abandoned mines in the outback of South Australia. Sprigg had been sent to find out whether some of the mines might be worth working again. He was several hundred miles from the nearest sea, in a remote area called the Ediacara Hills. Sprigg was eating his lunch, the story has it, when he turned over a rock and noticed what looked like some delicate fossils of jellyfish. As a geologist, he knew the rocks were so old that the finding was important. But he was not an established researcher of fossils, and when he wrote up his paper, few people took it seriously. The journal *Nature* rejected it, and Sprigg then worked his way down from journal to journal, until his article on what he called "Early Cambrian (?) Jellyfishes" appeared in the *Transactions of the Royal Society of South Australia* in 1947, alongside such papers as "On the Weights of Some Australian Mammals." The paper had a quiet career at first, and it took another decade or so before anyone realized what Sprigg had found.

At the time, scientists familiar with the fossil record were well aware of the importance of the Cambrian period, which began about 542 million years ago. In the "Cambrian explosion," a great range of the animal body plans we know today first appeared. Sprigg's discoveries turned out to be the first fossil record of animals living before that time. Sprigg did not realize this in 1947—he dated his jellyfish as early Cambrian. But as similar fossils were found in other places around the world and people took more note of Sprigg's outback jellyfish, it became clear that they dated from well before the Cambrian, and were probably not jellyfish, in most cases, at all. The period in prehistory now known as the *Ediacaran* (named after the hills Sprigg was exploring, and pronounced "Eedee-**ac**-aran") runs from 635 to about 542 million years ago. With the Ediacaran fossils we get our first direct evidence of what the lives of very early animals might have been like—how big they were, how numerous, how they lived.

The nearest large city to Sprigg's site is Adelaide, where a large collection of Ediacaran fossils is kept in the South Australian Museum. I was shown around the exhibits by Jim Gehling, who knew Sprigg and has worked on the fossils since 1972. I was surprised at how dense with life the ancient environment was; the Ediacaran was not about a few lone individuals. Many rock slabs Gehling has collected contain dozens of fossils of different sizes. Among the more prominent is *Dickinsonia*, which has fine stripe-like segments and looks a bit like a lily pad, or a bath mat. (A picture of a *Dickinsonia* in the South Australian Museum's collection appears just below this paragraph.) But if you focus on the large fossils, you miss most of the life that is present. Several times, Gehling walked up to what looked like a scrappy and nondescript bit of one of the rocks and pressed a piece of Silly Putty into it; when he took it back, the putty revealed a fine and detailed imprint of a tiny animal.

Ediacaran animals weren't tiny—many were several inches in length, some up to three feet. They seem to have mostly lived on

the sea floor, on and amid mats made of living material—clods
of bacteria and other microbes. Their world was a kind of
undersea swamp. Many were probably motionless as adults, anchored in place. Some might have been early sponges and corals.
Others had body forms that have since been entirely abandoned
by evolution—three-sided and four-sided designs, some with
quilted arrangements of plant-like fronds. Many Ediacarans seem
to have lived quiet lives of very limited mobility on the bottom of
the sea.

DNA evidence, though, suggests strongly that there were
nervous systems present at this time—probably in some of the animals on the wall in Adelaide. Which ones? Among them are some
animals who appear to have moved under their own steam. The
clearest case is *Kimberella*. This animal, which I have drawn below,
seems to have looked like the top half of a macaron, though a
macaron that was oval, with a front and back, and perhaps with
a tongue-like appendage on one end. The traces it left suggest
that it pushed the sediment before it as it moved, and scratched
the surfaces it crawled over, perhaps in feeding. *Kimberella* is
sometimes interpreted as a mollusk, or perhaps a member of an
abandoned evolutionary line close to the mollusks. If *Kimberella*
could crawl, then, especially as it grew to several inches long, it
almost certainly had a nervous system.

Kimberella seems the clearest case of a self-propelled Ediacaran, but there were very likely others. Near a *Dickinsonia*
fossil, one often finds a sequence of fainter traces bearing the
same shape. The animal seemed to sit and feed for a while at
one spot, then move on. Some reconstructions of Ediacaran

scenes show a few animals swimming, including *Spriggina*, named after Reg, their discoverer, but Gehling thinks this scenario is unlikely, because *Spriggina* fossils are always found the same way up. If a *Spriggina* swam, then whenever some tiny disaster killed it, it would have had some chance of landing the other way up. So Gehling thinks that *Spriggina*, like *Kimberella*, crawled.

Some biologists have argued that the Ediacarans are members of an animal-*like* evolutionary experiment, but not properly animals themselves. Rather than sitting on the animal branch of the tree of life, they exhibit a different way that cells can come together to yield an organism. Those strange three-sided forms and quilted fronds might support such a view. A more standard interpretation is that some Ediacarans, like *Kimberella*, were members of familiar animal groups, while other fossils represent abandoned evolutionary detours, together with ancient algae and other kinds of life. One theme that has emerged fairly consistently, though, is that the Ediacaran world was a rather *peaceful* one, a world largely without conflict and predation.

The word "peace" might not be apt, as it suggests a kind of considered friendship or truce. Rather, the Ediacarans appear to have had very little to *do* with each other. They munched on the mat, filtered food from the water, and in some cases roamed around, but if the fossil evidence is any guide, they hardly interacted at all.

Perhaps the fossil record is *not* a good guide; back in the first part of this chapter I discussed how the world of single-celled organisms now seems full of hidden interactions, mediated by chemical signals. The same may have been true in Ediacaran times, and this mode of interaction would leave no fossil trace. And certainly the Ediacarans competed with each other in an evolutionary sense—that is inevitable in a world of reproducing organisms. But some of the most conspicuous forms of interaction between one organism and another do seem to be absent. In particular, there is no evidence of predation—no half-eaten animal remains. (A few fossils show possible signs of predation-related damage in one animal, *Cloudina*, but even this case is unclear.) This was in no sense a dog-eat-dog world. Instead, in a phrase coined by the American paleontologist Mark McMenamin, it seems to have been "the Garden of Ediacara."

We can also learn something about life in the garden from Ediacaran bodies. These creatures do not seem to have large and complex sense organs. There are no large eyes, no antennae. Almost certainly they had some responsiveness to light and chemical traces, but they made little *investment*, as far as we can tell, in this sort of machinery. There are also no claws, spikes or shells—no weapons, and no shields with which to fend weapons off. Their lives seem not to have been lives of conflict and complicated interaction; they certainly didn't evolve the familiar tools used in such interactions. It was a garden of relatively self-contained and self-possessed beings. Macarons that pass in the night.

This is vastly unlike animal life now. Our animal cousins are highly alert to their environment; they track friends, foes, and countless other features of the landscape. They do that because what's going on around them *matters*; often it's a matter of life and death. Ediacaran lives show no evident sign of this moment-to-moment engagement with the environment. If so, this makes it likely that our Ediacaran ancestors put their nervous systems—when they had them—to different uses from those seen in more recent animals. Specifically, this might have been a time when the role played by those nervous systems fits the second of the theories of nervous system evolution I introduced above, the view based on internal coordination rather than sensory-motor control. Nervous systems were for shaping movements, maintaining rhythms, crawling and (perhaps) swimming. This would have included some sensing of the environment, but perhaps not very much.

Those inferences might be mistaken; perhaps a great deal of sensing and interaction was going on, using organs made of soft materials that leave no trace. Something else that has always puzzled me in discussions of the peaceful Ediacaran is the role of jellyfish. Sprigg's own fossils were not jellyfish, as he'd thought, but jellyfish are believed to have been around at this time, usually leaving no traces. Cnidarians in general, but especially jellyfish, have stinging cells, and a garden of stinging jellyfish, as any Australian will insist, is far from Edenic.

When the Royal Society of London held a conference on early animals and the first nervous systems in 2015, the age of the first jellyfish stings was a topic of puzzled discussion. It does seem that cnidarian stings evolved early—this we infer from the fact that the evolutionary split between two major branches of this group appears to date to the Ediacaran or even before, and animals on both sides of the split have the same sort of stingers. Cnidarian stings are *weapons*. Were they offensive or defensive? Neither the prey nor the foes of modern cnidarians existed back then. So who were the stings aimed at? We do not know.

Even if Ediacaran life was not as peaceful as has sometimes been supposed, a very different world was around the corner.

The "Cambrian explosion" began around 542 million years ago. In a relatively sudden series of events, most of the basic animal forms seen today arose. These "basic animal forms" did not include mammals, but did include vertebrates, in the form of fish. They also included arthropods—animals with an external skeleton and limbs with joints, such as trilobites—along with worms, and various others.

Why did it happen then, and why did it happen so fast? The

timing may have had to do with changes to the Earth's chemistry and climate. But the process itself may have been largely driven by a kind of evolutionary feedback, due to interactions between organisms themselves. In the Cambrian, animals became *part of each other's lives* in a new way, especially through predation. This means that when one kind of organism evolves a little, it changes the environment faced by other organisms, which evolve in response. From the early Cambrian onward there was definitely predation, together with everything that predation encourages: tracking, chasing, defending. When prey starts to hide or defend itself, predators improve their ability to track and subdue, leading in turn to better defenses on the prey side. An "arms race" has begun. From the early part of the Cambrian, the fossil record of animal bodies contains exactly what was *not* seen in the Ediacaran—eyes, antennae, and claws. The evolution of nervous systems was heading down a new path.

The revolution in behavior seen in the Cambrian also occurred, in large part, through the unfolding of possibilities inherent in a particular kind of *body*. A jellyfish has a top and bottom but no left and right. It is said to have radial symmetry. But humans, fish, octopuses, ants, and earthworms are all *bilaterians*, or bilaterally symmetrical animals. We have a front and back, and hence a left and right, as well as a top and bottom. The first bilaterians, or at least some early ones, might have looked bit like this:

I have given the animal eyespots on each side of its "head," though this is controversial (and those eyes are exaggerated in the picture—they would probably have been tiny). I am being generous to the early bilaterians.

Several Ediacaran animals are believed to be bilaterians,

including *Kimberella*, pictured a few pages back. If *Kimberella* was a bilaterian, then bilaterians before the Cambrian were already living somewhat more active lives than other animals. But in the Cambrian, they were unstoppable. The bilaterian body plan makes for mobility (walking is a very bilateral thing to do), and this body plan is friendly, it turns out, to many kinds of complex behavior. The diversification and entanglement of lives that took place in the Cambrian was mostly the work of bilaterians.

Before pressing on into the world of bilaterian evolution, let's pause and ask: which animal produces the most sophisticated behavior, which is the smartest, *without* a bilaterian body plan? Questions like this are notoriously hard to answer in an unbiased way, but in this case, the answer is clear. The most behaviorally sophisticated animals outside the bilaterians are the—terrifying—box jellyfish, the Cubozoa.

With their soft bodies and sparse fossil record, it is hard to work out when different kinds of jellyfish evolved, but cubozoans are thought to be late arrivals, originating in the Cambrian or after. A general feature of cnidarians, as I noted above, is their stinging cells. Some cubozoans have truly brutal venom in their stingers, strong enough to have killed large numbers of humans. In northeastern Australia, the presence of box jellyfish clears the beaches completely each summer; for a good part of the year it's too dangerous to swim off the shore at all, except in netted enclosures. To compound the problem, these jellyfish are invisible in the water. They also have the most complex behaviors of any non-bilaterian. Around the top of their body are two dozen sophisticated eyes—eyes with lenses and retinas, like ours. The Cubozoa can swim at about three knots, and some can navigate by watching external landmarks on the shore. Box jellyfish, the lethal behavioral pinnacle of non-bilaterian evolution, are also products of the new world that began in the Cambrian.

~ *Senses*

Nervous systems evolved before the bilaterian body plan, but this body created vast new possibilities for their use. During the Cambrian the relations between one animal and another became a more important factor in the lives of each. Behavior became *directed* on other animals—watching, seizing, and evading. From early in the Cambrian we see fossils that display the machinery of these interactions: eyes, claws, antennae. These animals also have obvious marks of mobility: legs and fins. Legs and fins don't necessarily show that one animal was interacting with others. Claws, in contrast, have little ambiguity.

In the Ediacaran, other animals might be there around you, without being especially relevant. In the Cambrian, each animal becomes an important part of the environment of others. This entanglement of one life in another, and its evolutionary consequences, is due to behavior and the mechanisms controlling it. *From this point on, the mind evolved in response to other minds.*

When I say that, you might reply that the term "mind" is out of place. In this chapter, I won't argue with that. Fine. What *is* the case, though, is that the senses, the nervous systems, and the behaviors of each animal began to evolve in response to the senses, nervous systems, and behaviors of others. The actions of one animal created opportunities for and demands on others. If a yard-long, fast-swimming *anomalocarid* is swooping down toward you, like a giant predatory cockroach with two grasping appendages on its head poised and ready, it's a very good thing to *know*, somehow, that this is happening, and to take evasive action.

The senses may well have been crucial to the Cambrian: organisms opened up to the world, especially to each other. The first sophisticated eyes seem to have appeared, eyes that can form an image. The Cambrian witnessed the appearance of both the *compound eyes* seen today in insects and *camera eyes* like our own. Imagine the behavioral and evolutionary consequences of being able to see the objects around you for the first time, especially objects at some distance and in motion. The biologist Andrew Parker has argued that the invention of eyes was *the* decisive event in the Cambrian. Others have developed broader views, but with a similar flavor. As the paleontologist Roy Plotnick and his colleagues put it, the result of this sensory opening was a "Cambrian information revolution." With an influx of sensory information comes a need for complex internal processing. When more is known, decisions become more complicated. (Is the anomalocarid more likely to intercept me if I flee to that hole, or that other one?) An image-forming eye makes possible actions that would be unthinkable without it.

Jim Gehling, my Ediacaran guide, and the British paleontologist Graham Budd have offered scenarios for how the feedback process generating these changes got under way. Near the close of the Ediacaran, Gehling suspects that scavenging arose, followed by predation. Animals went from feeding on microbial mats to feeding on the dead, and then began hunting the living. As Budd sees it, animal behavior itself changed the way resources were distributed in the Ediacaran. Imagine a world with edible microbial mats stretching before you like an endless swampy lawn. Slow-moving grazers wander over the mats, consuming this rather uniform resource. Other animals fed without moving. These animals then *become* a new kind of resource; they are big concentrations of nutritious carbon compounds. Nutrition is now less spread out than it was. It exists in patches. These animals might first have only been consumed by others after they had died. But this soon changed. Scavenging became predation.

If the fossil record is taken at face value, it seems that one group set the pace: the *arthropods*. This group today includes insects, crabs, and spiders. Early in the Cambrian we see the rise of *trilobites*, which are prototypical arthropods with shells, jointed legs, and compound eyes. In the photograph of the *Dickinsonia* fossil on page 29, you'll find two much smaller fossils just below it, above the letters "A" and "B." These animals are just millimeters long, and Gehling thinks they might be precursors of trilobites—still soft-bodied, but with hints of a trilobite design. In this picture, *Dickinsonia* is present in its classic Ediacaran mode, with no apparent limbs, head, or protection, while purposeful little bugs lurk beneath. The image reminds me of a drawing in a book about the dinosaurs and their decline that I owned as a child. A huge dinosaur towered over a few small and mischievous-looking mammals, shrew-like creatures, at its feet. I think they had their eye on a clutch of dinosaur eggs. The trilobite precursors look intent on a similar goal, with the lilypad-bathmat *Dickinsonia* oblivious above.

Michael Trestman, another philosopher, has offered an interesting way of looking at all these animals. Consider, he says, the category of animals who have *complex active bodies*. These are animals who can move quickly, and who can seize and manipulate objects. Their bodies have appendages that can move in many directions, and they have senses, such as eyes, which can track distant objects. Trestman says that only three of the major animal groups produced some species with complex active bodies (CABs). Those groups are arthropods, *chordates* (animals like us with a nerve cord down their back), and one group of mollusks, the cephalopods. This trio might seem to make up a large category, because these are the sorts of animals that tend to come to our minds, but it is a small group in many ways. There are about thirty-four animal *phyla*—basic animal body plans. Only three phyla contain some animals with CABs, and within one of those three, the mollusks, the only animals that count are cephalopods.

With these ancient stages of the historical story in place, I'll return to the divide between two views of nervous systems and their evolution—the sensory-motor and action-shaping views. Earlier I introduced the distinction, linked it to two roles that signals can have in social life (sexton and Revere versus the rowboat), and noted that the two roles are different but also compatible. What might be the historical significance of this divide? Can the distinction be fit in some natural way onto the march of millennia from the Ediacaran, to the Cambrian, to more recent times? It does seem possible that there was a shift in the roles nervous systems were performing. Although tracking events in the outside world might always be worth doing to some extent, the Cambrian sees a great increase in the importance of this side of life. There's more that's worth watching, and more that needs to be done in response to what's seen. Not paying attention, for the first time, means getting eaten by the swooping anomalocarid. Perhaps, then, the very first nervous systems primarily served to coordinate actions—first animating the body of an ancient cnidarian, then shaping the actions of Ediacarans. But if there was such an era, by the Cambrian it was over.

This is one possibility among many, though, and our imaginations, shaped by lives lived in modern bodies, underestimate the range of options. Possibilities abound. Here is one developed by the biologist Detlev Arendt and his colleagues. As they see it, nervous systems originated twice. But they don't mean that they evolved in two kinds of animals; rather, they originated twice in the *same* animals, at different places in the animal's body. Imagine a jellyfish-like animal shaped like a dome, with a mouth underneath. One nervous system evolves on the top, and tracks light, but not as a guide to action. Instead it uses light to control bodily rhythms and regulate hormones. Another nervous system evolves to control movement, initially just the movement of the mouth. And at some stage, the two systems begin to move within the body, coming into new relations with each other. Arendt sees this as one of the crucial events that took bilaterians forward in the Cambrian. A part of the body-controlling system moved up toward the top of the animal, where the light-sensitive system sat. This light-sensitive system, again, was only guiding chemical changes and cycles, not behavior. But the joining of the two nervous systems gave them a new role.

What an amazing image: in a long evolutionary process, a motion-controlling brain marches up through your head to meet there some light-sensitive organs, which become eyes.

~ The Fork

The bilaterian body plan arose before the Cambrian, in some small and unremarkable form, but it became the bodily scaffold on which a long series of increases in behavioral complexity was laid down. Early bilaterians also have another role in this book. Sometime soon after they appeared, probably still in the Ediacaran, there was a branching, one of the countless evolutionary forks that take place as the millennia pass. A population of these animals split into two. The animals who initially wandered off down the two paths might have looked like small flattened worms. They had neurons, and perhaps very simple eyes, but little of the complexity that was to come. Their scale was measured perhaps in millimeters.

After this innocuous split, the animals on each side diverged, and each became ancestor to a huge and persisting branch of the tree of life. One side led to a group that includes vertebrates, along with some surprising companions such as starfish, while the second side led to a huge range of other invertebrate animals. The point just before this split is the last point at which an evolutionary history is *shared* between ourselves and the big group of invertebrates that includes beetles, lobsters, slugs, ants, and moths.

Here is a diagram of this part of the tree of life. Lots of groups are omitted from the picture, both outside and inside the branches shown. The moment we're talking about is labeled "the fork."

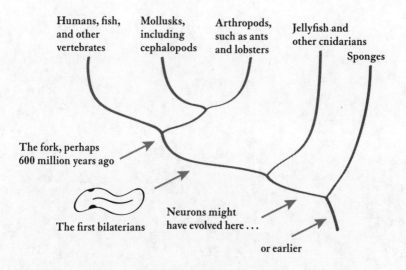

On each path downstream of the fork, more branchings occurred. One side eventually sees fish appear, then dinosaurs and mammals. This is our side. On the other side, further branchings give rise to arthropods, mollusks, and others. On *both* sides, passing from the Ediacaran into the Cambrian and beyond, lives become entangled, the senses open, and nervous systems expand. Until, in one tiny example of this sensory and behavioral entangling, a rubber-encased mammal and a color-changing cephalopod find themselves staring at each other in the Pacific Ocean.

3

MISCHIEF AND CRAFT

Mischief and craft are plainly seen to be characteristics of this creature.
—Claudius Aelianus, third century A.D.,
writing about the octopus

In a Sponge Garden

Someone is watching you, intently, but you can't see them. Then you notice, drawn somehow by their eyes.

You're amid a sponge garden, the sea floor scattered with shrub-like clumps of bright orange sponge. Tangled in one of these sponges, and the gray-green seaweed around it, is an animal about the size of a cat. Its body, though, seems to be everywhere and nowhere. Much of the animal seems to have no definite shape at all. The only parts you can keep a fix on are a small head and the two eyes. As you make your way around the sponge, so too do those eyes, keeping their distance, keeping part of the sponge between the two of you. Its color matches—exactly, perfectly—the seaweed around it, except that some of its skin is folded into tiny tower-like peaks, and the tips of these peaks match—nearly as exactly—the orange of the sponge. You keep coming round its side of the sponge, and eventually it raises its head high, then rockets away under jet propulsion.

A second meeting with an octopus: this one is in a den. Shells are strewn in front, arranged with some pieces of old glass. You stop in front of its house and the two of you look at each other. This one is small, about the size of a tennis ball. You reach forward a hand and stretch out one finger, and one octopus arm slowly uncoils and comes out to touch you. The suckers grab your skin, and the hold is disconcertingly tight. Having attached the suckers, it tugs your finger, pulling you gently in. The arm is packed with sensors, hundreds of them in each of the dozens of suckers. It's *tasting* your finger as it draws it in. The arm itself is alive with neurons, a nest of nervous activity. Behind the arm, large round eyes watch you the whole time. Hundreds of millions of years on from the events of chapter 2, this is one place the evolution of animals has landed.

~ Evolution of the Cephalopods

Octopuses and other cephalopods are *mollusks*—they belong to a large group of animals which also includes clams, oysters, and snails. Part of the story of the octopus, then, is the evolutionary history of mollusks. In the previous chapter we reached the Cambrian, the period in the history of life when a great range of animal body plans appear in the fossil record. Many of these animal groups, including mollusks, must pre-date the Cambrian, but in the Cambrian mollusks become noticeable, because of their shells.

Shells were the mollusks' response to what looks like an abrupt change in the lives of animals: the invention of predation.

There are various ways of dealing with the fact that you are suddenly surrounded by creatures who can see and would like to eat you, but one way, a molluscan specialty, is to grow a hard shell and live within or beneath it. The cephalopod line probably goes back to an early mollusk of this kind, crawling along the bottom of the sea under a hard shell peaked like a cap. This animal looked a bit like a limpet, one of those plain, cup-like shellfish that grip rocks in tide pools today. The cap grew, Pinocchio-like, over evolutionary time, slowly taking the shape of a horn. These animals were small—the "horn" was less than an inch long. Beneath the shell, as with other mollusks, a muscular "foot" anchored the animal and enabled it to crawl along the sea floor.

Then, at a stage later in the Cambrian, some of these animals rose from the sea floor and entered the water column. On dry land, no effortless move up into the air is possible for an animal; such a move requires the expense of wings or something similar. In the sea you can lift off easily, be carried, and see where you end up.

An upward-pointing shell which protects can be made into a buoyancy device, by filling it with gas. Early cephalopods seem to have done just that. Making the shell buoyant may have initially enabled easier crawling, and many of the old cephalopods might have moved by engaging in a half-crawl, half-swim on the bottom of the sea. Some, though, rose higher, and found a world of opportunity above. A small amount of gas, held within the shell, will turn a limpet into a zeppelin.

Once aloft, the "foot" is no use for crawling, so the zeppelin-cephalopods invented jet propulsion, by directing water through a tube-like *siphon* which could be pointed in several directions. The foot itself was freed up for grasping and manipulating objects, and part of it flowered into a cluster of tentacles. Talk of "flowering" would sound inappropriate, though, to the animals on the other end of these tentacles—the animals being grasped—as some of the tentacles sprouted dozens of sharp hooks. The opportunity the cephalopods were seizing by rising up into the water was the opportunity to feed on other animals, to become predators themselves. This they did with great evolutionary enthusiasm. Many forms appeared, with straight shells and coiled, and the largest reached sizes of eighteen feet or more. Beginning as diminutive limpets, cephalopods had become the most fearsome predators in the sea.

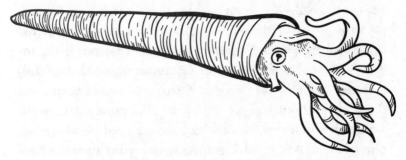

As well as zeppelins, a range of cephalopod hovercrafts and tanks probably prowled the sea floor—some of the shells from this time seem too unwieldy to carry in the open water. All these animals are now extinct, with one non-fearsome exception, the nautilus. Many of the losses occurred as part of the mass extinctions that punctuate the history of life, but it's also likely that some predatory cephalopods were slowly outcompeted by fish, as those fish became larger and better armed. The zeppelins were challenged, and eventually vanquished, by airplanes.

The nautilus, however, made it through. No one knows why. At the start of this book I cited a Hawaiian creation myth that judges the octopus a "lone survivor" from an earlier world. The real survivor is indeed a cephalopod, but nautilus rather than octopus. Still living in the Pacific, present-day nautiluses are little changed from 200 million years ago. Living in coiled shells, they're now scavengers. They have simple eyes and a cluster of tentacles, and they move up and down, from the deep sea to shallower water, in a rhythm that's still being studied. They seem to stay higher in the water at night, deeper in the day.

Another shift was to come in the evolution of cephalopod bodies. Sometime before the age of the dinosaurs, it seems, some cephalopods began to give up their shells. The protective casings that had become buoyancy devices were abandoned, reduced, or internalized. This enabled more freedom of movement, but at the price of greatly increased vulnerability. It seems quite a gamble, but this was a path taken several times. The last common ancestor of "modern" cephalopods is not known, but at some stage the lineage split into two main branches, an eight-armed group including octopuses and a ten-armed group including cuttlefish and squid. These animals reduced their shells in different ways. In the cuttlefish, a shell was retained internally, and still helps the animal remain buoyant. In squid, a sword-shaped internal structure called a "pen" remains. Octopuses have lost their shell entirely. Many cephalopods began to live as soft-bodied, unprotected animals on reefs in shallow seas.

The oldest *possible* octopus fossil dates from 290 million years ago. I emphasize the uncertainty—it's just one specimen, and little more than a smudge on a rock. After this there is a gap in the record, and then at around 164 million years ago there is a clearer case, a fossil that looks undeniably like an octopus, with eight arms and an octopus-like pose. The fossil record of octopuses remains skimpy because they don't preserve well. But at some stage they radiated; around 300 species are known at present, including deep-sea as well as reef-dwelling forms. They range from less than an inch in length to the giant Pacific octopus, which weighs in at 100 pounds and spans twenty feet from arm tip to arm tip.

That's the journey of the cephalopod body, a path from Ediacaran macaron through limpet-like shellfish to predatory hovercraft and zeppelin. The encumbrance of the external shell is then abandoned, as the shell is brought inside the body or, in an octopus, lost completely. With that step, the octopus loses almost all definite shape.

To completely forgo both skeleton and shell is an unusual evolutionary move for a creature of this size and complexity. An octopus has almost no hard parts at all—its eyes and beak are the largest—and as a result it can squeeze through a hole about the size of its eyeball and transform its body shape almost indefinitely. The evolution of cephalopods yielded, in the octopus, a body of pure possibility.

During the time I was writing an early version of this chapter, I spent a few days watching a pair of octopuses in the rocky shallows. I saw them mate once, and then spend much of the next afternoon just sitting, it seemed. The female moved off a little way, but returned to her den as the sun got low. The male had spent the day in a more exposed spot, less than a foot from her den. He was there when she came back.

I watched them, off and on, for two afternoons, and then storms came. Winds of sixty miles per hour lashed the coast, and waves rolled in from the south. The bay where the octopuses live has some protection from this onslaught, but not much. Waves swept around the entrance and turned the water into a boiling white soup. The shore was beaten by these storms for the next four days. Where do the octopuses go when the waves are pounding their rocks? It was impossible to get into the water to see. The cuttlefish have no problem. They disappear for weeks when the weather is bad. They fire up their jet propulsion and move off to some unknown deeper place. Perhaps the octopuses also move further out to sea, but more likely they climb into a crevice and hang on, for days at a stretch, recalling their ancestors who gripped rocks from inside cap-shaped shells.

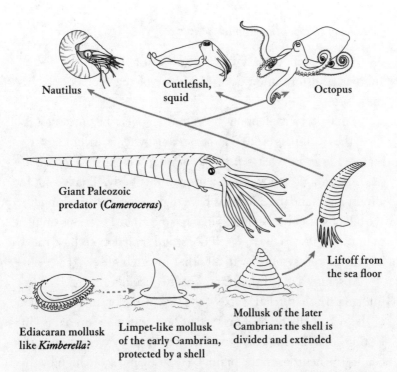

Nautilus

Cuttlefish, squid

Octopus

Giant Paleozoic predator (*Cameroceras*)

Liftoff from the sea floor

Ediacaran mollusk like *Kimberella*?

Limpet-like mollusk of the early Cambrian, protected by a shell

Mollusk of the later Cambrian: the shell is divided and extended

Evolution of the Cephalopods: The figure is not to scale (far from it), and doesn't represent actual descent relations between species. It presents a chronological sequence of forms seen in cephalopod evolution from over half a billion years ago to the present, with a few of the most important branchings marked along the way. I have included the controversial *Kimberella* as a possible early stage. The capped limpet-like shellfish is a monoplacophoran. The next animal, with a shell divided into compartments, is something like *Tannuella*. Opinion seems divided on whether the next in line, *Plectronoceras*, had lifted off the ground or was still on the sea floor, but this animal is often regarded as the first "true" cephalopod, because of various internal features. *Cameroceras* is the giant of the large predatory cephalopods, with conservative length estimates of up to eighteen feet. The octopus and squid are descended from unknown cephalopods that gave up their external shells and are now extinct, unlike the nautilus, which kept its shell and lived on.

~ *Puzzles of Octopus Intelligence*

As the cephalopod body evolved toward its present-day forms, another transformation occurred: some of the cephalopods became smart.

"Smart" is a contentious term to use, so let's begin cautiously. First, these animals evolved large nervous systems, including large brains. Large in what sense? A common octopus (*Octopus vulgaris*) has about 500 million neurons in its body. That's a lot by almost any standard. Humans have many more—something like 100 billion—but the octopus is in the same range as various smaller mammals, close to the range of dogs, and cephalopods have much larger nervous systems than all other invertebrates.

Absolute size is important, but it is usually regarded as less informative than relative size—the size of the brain as a fraction of the size of the body. This tells us how much an animal is "investing" in its brain. This comparison is made by weight, and only counts the neurons in the brain. Octopuses also score high by this measure, roughly in the range of vertebrates, though not as high as mammals. Biologists regard all these assessments of size, though, as only a very rough guide to the brain*power* an animal has. Some brains are organized differently from others, with more or fewer synapses, and those synapses can also be more or less complicated. The most startling finding in recent work on animal intelligence is how smart some birds are, especially parrots and crows. Birds have quite small brains in absolute terms, but very high-powered ones.

When we try to compare one animal's brainpower with another's, we also run into the fact that there is no single scale on which intelligence can be sensibly measured. Different animals are good at different things, as makes sense given the different lives they live. An analogy can be drawn with tool kits: brains are like tool kits for the control of behavior. As with human tool kits, there are some elements in common across many trades, but much diversity also. All the tool kits found in animals include some kind of perception, though different animals have very different ways of taking in information. All (or almost all) bilaterian animals have some form of memory and a means for learning, enabling past experiences to be brought to bear on the present. The tool kit sometimes includes capacities for problem solving and planning. Some tool kits are more elaborate and expensive than others, but they can be sophisticated in different ways. One animal might have better senses, while another may have more sophisticated learning. Different tool kits go with different ways of making a living.

When comparing cephalopods with mammals, the difficulties are acute. Octopuses and other cephalopods have exceptionally good eyes, and these are eyes built on the same general design as ours. Two experiments in the evolution of large nervous systems landed on similar ways of seeing. But the nervous systems beneath those eyes are organized very differently. When biologists look at a bird, a mammal, even a fish, they are able to map many parts of one animal's brain onto another's. Vertebrate brains all have a common architecture. When vertebrate brains are compared to octopus brains, all bets—or rather, all mappings—are off. There is no part-by-part correspondence between the parts of their brains and ours. Indeed, octopuses have not even collected the majority of their neurons inside their brains; most of the neurons are found in their arms. Given all this, the way to work out how smart octopuses are is to look at what they can *do.*

Here we quickly encounter puzzles. Perhaps the heart of the matter is a mismatch between the results of laboratory experiments on learning and intelligence, on one side, and a range of anecdotes and one-off reports, on the other. Mismatches like this are common in the world of animal psychology, but they are especially acute in the case of octopuses.

When tested in the lab, octopuses have done fairly well, without showing themselves to be Einsteins. They can learn to navigate simple mazes. They can use visual cues to determine which of two possible environments they have been placed in, and then take the correct route to a goal for that environment. They can learn to unscrew jars to obtain the food inside. But octopuses are slow learners in all these contexts. When you read the fine print of a "successful" experiment, progress often seems agonizingly slow. Against a background of mixed experimental results, though, there are anecdotes suggesting that a lot more is going on. What I find most intriguing is the octopus's ability to adapt to new and unusual circumstances—confinement in a lab—and turn the apparatus around them to their own octopodean purposes.

A lot of early octopus work was done in Italy, at the Naples Zoological Station, in the middle of the twentieth century. Peter Dews was a Harvard scientist who worked mostly on the interaction between drugs and behavior. He had a general interest in learning, though, and his octopus experiment did not involve drugs at all. Dews was influenced by his Harvard colleague B. F. Skinner, whose work on "operant conditioning"—the learning of behaviors by reward and punishment—had revolutionized psychology. The idea that successful behaviors will be repeated and unsuccessful ones abandoned had been pioneered by Edward Thorndike around 1900, but Skinner developed the idea in great detail. Dews, with many others, was inspired by the way Skinner was able to make animal experiments rigorous and exact.

In 1959 Dews applied some standard experiments on learning and reinforcement to octopuses. Octopuses may be distantly related to vertebrates like us, but do they learn in similar ways? Can they learn, for example, that pulling and releasing a lever will get them a reward, and come to produce this behavior at will?

I first came across Dews's work through a brief mention of his experiment in Roger Hanlon and John Messenger's book *Cephalopod Behaviour*. Hanlon and Messenger comment that pulling and releasing a lever is surely something an octopus would never do in the sea, and they say that Dews's experiment was not successful. I was curious about how things went, though, so I went back to the 1959 paper. The first thing I noticed is that the experiment *was* successful with respect to its main goals. Dews trained three octopuses, and found that all three of them did learn to operate the lever to obtain food. When they pulled the lever, a light came on and a small piece of sardine was given as a reward. Two of the octopuses, named Albert and Bertram, did this in a "reasonably consistent" manner, Dews said. The behavior of the third octopus, named Charles, was different. Though Charles did pass the test in a minimal way, his handling of the situation encapsulates much of the story with octopus behavior. Dews wrote:

1. Whereas Albert and Bertram gently operated the lever while free-floating, Charles anchored several tentacles on the side of the tank and others around the lever and applied great force. The lever was bent a number of times, and on the 11th day was broken, leading to a premature termination of the experiment.
2. The light, suspended a little above the level of the water, was not the subject of much "attention" by Albert or Bertram; but Charles repeatedly encircled the lamp with tentacles and applied considerable

force, tending to carry the light into the tank. This behavior is obviously incompatible with lever-pulling behavior.

3. Charles had a high tendency to direct jets of water out of the tank; specifically, they were in the direction of the experimenter. The animal spent much time with eyes above the surface of the water, directing a jet of water at any individual who approached the tank. This behavior interfered materially with the smooth conduct of the experiments, and is, again, clearly incompatible with lever-pulling.

Dews comments dryly, "The variables responsible for the maintenance and strengthening of the lamp-pulling and squirting behavior in this animal were not apparent." The language Dews is using here—the language of "variables responsible," and so on—shows that he is thinking (or writing, at least) in line with the assumptions of mid-twentieth-century animal behavior experiments. He assumes that if Charles is squirting experimenters and absconding with the apparatus, this must be because something in Charles's history has reinforced this behavior. Animals of a given species will start out the same, on this view, and if they diverge in behavior this must be because of rewarding (or unrewarding) experiences. That is the framework Dews is working within. However, one message of octopus experiments is that there is a great deal of individual variability. Charles, most likely, was not an octopus who started with the same behavioral routines as the others and was reinforced for squirting experimenters, but an octopus with a particularly feisty temperament.

This 1959 paper was one of the first encounters between a tightly controlled style of scientific work on animal behavior and the idiosyncrasies of the octopus. A great deal of work on animals has been done under the assumption that all animals of a given species (and perhaps of a given sex) will be very similar until they encounter different rewards, and will peck or run or pull a lever all day in order to get the same little morsels of food. Dews, like many others, wanted to work this way because he was determined to use what he called "objective, quantitative methods of study." I am all for those, too. But octopuses, far more than rats and pigeons, have their own ideas: "mischief and craft," as Aelianus, in this chapter's epigraph, had it.

The most famous octopus anecdotes are tales of escape and thievery, in which octopuses in aquariums raid neighboring tanks at night for food. Those stories, despite their charm, are not especially indicative of high intelligence. Neighboring tanks are not so different from tide pools, even though the entrance and exit take more effort. Here is a behavior I find more intriguing. Octopuses in at least two aquariums have learned to turn off the lights by squirting jets of water at the bulbs when no one is watching, and short-circuiting the power supply. At the University of Otago in New Zealand, this became so expensive that the octopus had to be released back to the wild. A lab in Germany had the same problem. This seems very smart indeed. However, one can also sketch an explanation which may partially deflate the story. Octopuses don't like bright lights, and they squirt jets of water at all sorts of things that annoy them (as Peter Dews discovered). So squirting water at lights might not be something that requires much explanation. Also, octopuses are more likely to roam far enough away from their dens to squirt at this particular target when no humans are around. On the other hand, both the stories of this kind that I've seen give the impression that the octopus learned *very quickly* how well this behavior works—that it's worth getting into position and aiming right at the light, to turn it out. It should be possible to set up an experiment that tests some of the various possible explanations for the behavior.

This case illustrates a more general fact: octopuses have an ability to adapt to the special circumstances of captivity and their interaction with human keepers. Octopuses in the wild are fairly solitary animals. Their social life, in most species, is thought to be minimal (though later I'll look at exceptions to this pattern). In the lab, however, they are often quick to get the hang of how life works in their new circumstances. For example, it has long appeared that captive octopuses can recognize and behave differently toward individual human keepers. Stories of this kind have been coming out of different labs for years. Initially it all seemed anecdotal. In the same lab in New Zealand that had the "lights-out" problem, an octopus took a dislike to one member of the lab staff, for no obvious reason, and whenever that person passed by on the walkway behind the tank she received a jet of half a gallon of water in the back of her neck. Shelley Adamo, of Dalhousie University, had one cuttlefish who reliably squirted streams of water at all *new* visitors to the lab, and not at people who were often around. In 2010, an experiment confirmed that giant Pacific octopuses can indeed recognize individual humans, and can do this even when the humans are wearing identical uniforms.

Stefan Linquist, a philosopher who once studied octopus behavior in the lab, puts it like this: "When you work with fish, they have no idea they are in a tank, somewhere unnatural. With octopuses it is totally different. They know that they are inside this special place, and you are outside it. All their behaviors are affected by their awareness of captivity." Linquist's octopuses would mess around with their tank, manipulating and testing it. Linquist had a problem with octopuses deliberately plugging the outflow valves on the tanks by poking in their arms, perhaps to increase the water level. Of course, this flooded the entire lab.

Another tale that illustrates Linquist's point was told to me by Jean Boal, of Millersville University in Pennsylvania. Boal has a reputation as one of the most rigorous and critical of ceph-alopod researchers. She is known for her meticulous experimental designs, and her insistence that "cognition" or "thought" in these animals should be hypothesized only when experimental results cannot be explained in any simpler way. But like many research-ers, she has a few tales of behaviors that are baffling in what they seem to show about the inner lives of these animals. One of these incidents has stayed in her mind for over a decade. Octopuses love to eat crabs, but in the lab they are often fed on thawed-out frozen shrimp or squid. It takes octopuses a while to get used to these second-rate foods, but eventually they do. One day Boal was walking down a row of tanks, feeding each octopus a piece of thawed squid as she passed. On reaching the end of the row, she walked back the way she'd come. The octopus in the first tank, though, seemed to be waiting for her. It had not eaten its squid, but instead was holding it conspicuously. As Boal stood there, the octopus made its way slowly across the tank toward the outflow pipe, watching her all the way. When it reached the outflow pipe, still watching her, it dumped the scrap of squid down the drain.

This story, along with all the tales of octopuses squirting experimenters, reminded me of something I'd seen myself. Cap-tive octopuses often try to escape, and when they do, they seem unerringly able to pick the one moment you aren't watching them. If you have an octopus in a bucket of water, for example, it will often look content enough in there, but if your attention strays for a second, when you look back there will be an octopus qui-etly crawling across the floor.

I thought I might be imagining this tendency, until I heard a talk a few years ago given by David Scheel, who works with octopuses full-time. He, too, said that octopuses seem to track in subtle ways whether he is watching them or not, and they make their move when he isn't. I suppose this makes sense as a natural behavior in octopuses; you want to make a run for it when the barracuda is not looking at you, rather than when he is. But the fact that octopuses can so quickly do this with humans—both with scuba mask and without—is impressive.

As stories of this kind accumulate, an explanation suggests itself for the mixed results with octopuses in some standard learning experiments. It's often said that they don't do especially well in these experiments because the behaviors required are unnatural. (Hanlon and Messenger said this about the Dews experiment with the lever pulling, for example.) But octopus behavior in laboratory settings indicates that "unnatural" is often no problem for them. Octopuses can open screw-cap jars for food, and one has even been filmed opening such a jar from the inside. Behaviors don't get much more unnatural than that. I think the problems with the old Peter Dews experiment, such as they were, came in part from the assumption that an octopus would be *interested* in pulling a lever repeatedly to get pieces of sardine, collecting piece after piece of this second-rate food. Rats and pigeons will do things like that, but octopuses take a while to deal with each item of food, probably can't cram themselves, and tend to lose interest. For at least some of them, taking the lamp down from above the tank and hauling it back to the den—*that* is more interesting. So is squirting the experimenters.

In response to the difficulty of motivating the animals, some researchers, regrettably, have used negative reinforcement—electric shocks—more freely than they would with other animals. Quite a lot of the early work done in the Naples Zoological Station treated octopuses badly. Not only were electric shocks used, but many experiments included the removal of parts of the octopus's brain, or the cutting of important nerves, just to see what the octopus would do when it woke up. Until recently, octopuses could also be operated on without anesthetic. As invertebrates, they were not covered by animal cruelty rules. Many of these early experiments make for distressing reading for someone who regards octopuses as sentient beings. Over the last decade, however, octopuses have often been listed as a kind of "honorary vertebrate" in rules governing their treatment in experiments, especially in the European Union. This is a step forward.

Another octopus behavior that has made its way from anecdote to experimental investigation is *play*—interacting with objects just for the sake of it. An innovator in cephalopod research, Jennifer Mather, along with Roland Anderson of the Seattle Aquarium, did the first studies of this behavior, and it's now been investigated in detail. Some individual octopuses—and only some—will spend time blowing pill bottles around their tank with their jet, "bouncing" the bottle back and forth on the stream of water coming from the tank's intake valve. In general, the initial interest an octopus takes in any new object is gustatory—can I eat it? But once an object is found to be inedible, that does not always mean it's uninteresting. Recent work in the lab by Michael Kuba has confirmed that octopuses can quickly tell that some items are not food, and are often still quite interested in exploring and manipulating them.

~ Visiting Octopolis

In the first chapter I described Matthew Lawrence's discovery of an octopus site on the east coast of Australia. Matt explored the bay by dropping an anchor off his small boat, swimming down to pick it up, and letting the drift of the boat guide his wandering over the sea floor. (I should add that diving alone is a bad idea. Matt takes down a second air supply that is completely independent of the first, in case things go wrong. Even then, it's not recommended.) In 2009 he came across a shell bed with about a dozen octopuses living on it. They seemed unconcerned by his presence, roaming and wrestling with each other as he watched.

Matt marked the GPS coordinates of the spot and began visiting regularly. He'd watch and interact with the octopuses. They didn't seem to mind his presence at all, and some were curious enough to play with him and explore his equipment. His camera and air hoses soon had octopuses roaming over them. Others were too busy dealing with each other. Sometimes he saw what looked like a kind of "bullying" behavior. An octopus would be sitting quietly in its den, and a larger one would come over, jump on top of the den, and wrestle furiously with the one below. After a great multicolored convulsion, the octopus below would come flying out like a rocket, its body pale, and land a few meters away, just off the shell bed. The aggressor octopus would wander back to its den.

As time passed, Matt became more and more accustomed to dealing with these animals, and to this day it seems to me that the octopuses treat Matt differently from anyone else. Once at a site close to this one, an octopus grabbed his hand and walked off with him in tow. Matt followed, as if he were being led across the sea floor by a very small eight-legged child. The tour went on for ten minutes, and ended at the octopus's den.

Though he's not a biologist, Matt had a sense that his site might be unusual. He posted some photos on a website that functions as an information center for cephalopod-inclined hobbyists and scientists. There they were seen by the biologist Christine Huffard, who asked me: Did I know this place? I was startled when I read about what he'd found, and Matt's site is only a few hours from Sydney. I got in touch when I was next in town, and drove down to meet him.

Matt, I found, is a scuba fanatic. He keeps his own air compressor in a garage, where he concocts personalized mixes of enriched air to fill his tanks. Soon we were chugging out on his small boat to a spot in the middle of his bay, where he set the anchor and we swam down the line, observed by just a few small fish.

The site we now call Octopolis is about fifty feet down. It's almost invisible until you get quite close, and the sea floor around it is nondescript. Scallops live scattered in little clumps, or on their own, and various kinds of seaweed waft about on the sand. My first trip to the site, in cold winter water, was quiet. We found just four octopuses, who were not doing much. But I could tell it was an unusual place. There was a bed of scallop shells, as Matt had said, a couple of yards in diameter. It seemed to contain shells of many ages. An encrusted rock-like object, a foot high or so, sat in the middle, with the largest octopus on the site using it as a den. I took measurements and photos, and began coming back whenever I could. Soon I was seeing the high concentrations of octopuses and complex behaviors that Matt had encountered on his first dives there.

If we had air enough and time, I don't know how long we'd stay down there. When the site is active, it's enthralling. The octopuses eye each other from their dens among the shells. They periodically haul themselves out and move over the shell bed or away onto the sand. Some will pass by others without incident, but an octopus might also send out an arm to poke or probe at another. An arm, or two, might come back in response, and this leads sometimes to a settling-down, with each octopus going on its way, but in other cases it prompts a wrestling match.

The first photo on the next page was taken just off the edge of the site, and it's to give you a sense of how these animals look. The species is *Octopus tetricus*, a medium-size octopus found just in Australia and New Zealand. This is a fairly large individual; from the sea floor to the high spot at the end of its back would be a bit under two feet. It is rushing toward another octopus, off to the right.

The next scene is on the shell bed itself. The octopus on the left is leaping toward the one on the right, who is stretched out and starting to flee.

And this is a more serious fight, on the sand just off the edge of the site:

In order to study changes in the shell bed, I once brought out some stakes and hammered them into the sea floor to mark the site's approximate boundaries. The stakes, about seven inches long, were made of plastic, so I taped a heavy metal bolt to each one to give it more weight. I drove the stakes in so that only an inch or so of each one sat above the sand, and placed them at the four compass points. They're very inconspicuous, hard to see unless you know exactly where to look. Some months later I went out to the site again, and found that one of the stakes had been hauled out and added to the pile of debris around one of the octopus dens, some distance away. The stake, I think, would have quickly been found inedible, and it was probably not especially useful as a barricade. But as with tape measures, cameras, and many other things we bring down to the site, the stake's novelty seemed to make it interesting to an octopus.

Other octopus manipulations of foreign objects are done for more practical reasons. In 2009, a group of researchers in Indonesia were surprised to see octopuses in the wild carrying around pairs of half coconut shells to use as portable shelters. The shells, neatly halved, must have been cut by humans and discarded. The octopuses put them to good use. One half-shell would be nested inside another, and the octopus would carry the pair beneath its body as it "stilt-walked" across the sea bottom. The octopus would then assemble the halves into a sphere with itself inside. A wide range of animals use found objects for shelters (hermit crabs are an example), and some use tools for collecting food (including chimps and some crows). But to assemble and disassemble a "compound" object like this, and put it to use, is very rare. It's not clear what to compare this behavior to, in fact. Many animals combine a variety of materials when making nests—a lot of nests are "compound" objects. But those are not disassembled, carried around, and put back together.

The coconut-house behavior illustrates what I see as the distinctive feature of octopus intelligence; it makes clear the *way* they have become smart animals. They are smart in the sense of being curious and flexible; they are adventurous, opportunistic. With this idea on the table I can add more to my picture of how octopuses fit into the range of animals and the history of life.

In the previous chapter, using some ideas from Michael Trestman, I said that across the wide range of animal body plans, only three groups contain some species with "complex active bodies." Those are chordates (like us), arthropods (like insects and crabs), and a small group of mollusks, the cephalopods. The arthropods went down this road first, in the early Cambrian, over 500 million years ago. The way they did this may have initiated a process of evolutionary feedback that soon encompassed everyone else. Arthropods were first, and chordates and cephalopods followed.

Setting aside our own case, we can see a difference in the paths taken by the two other groups. Many arthropods specialize in social living and coordination. Not all of them do this—indeed, the majority of arthropod species don't—but in the area of behavior, many of the great arthropod achievements are social. This is seen especially in ant and honeybee colonies, and in the air-conditioned cities built by termites.

Cephalopods are different. They never went onto land (though some other mollusks did), and while they probably started on the road toward complex behavior at a later date than the arthropods, they eventually evolved larger brains. (Here I think of an ant colony as many organisms with many brains, not as one.) In arthropods, very complex behaviors tend to be achieved through the coordination of many individuals. Some squid are social, but with nothing like the organization of ants and honeybees. Cephalopods, with the partial exception of squid, acquired a non-social form of intelligence. The octopus, most of all, would follow a path of lone idiosyncratic complexity.

~ *Nervous Evolution*

Let's look more closely now at what's inside an octopus, and how the nervous system behind these behaviors evolved.

The history of large brains has, very roughly, the shape of a letter **Y**. At the branching center of the **Y** is the last common ancestor of vertebrates and mollusks. From here, many paths run forward, but I single out two of them, one leading to us and one to cephalopods. What features were present at that early stage, available to be carried forward down both paths? The ancestor at the center of the **Y** certainly had neurons. It was probably a worm-like creature with a simple nervous system, though. It may have had simple eyes. Its neurons may have been partly bunched together at its front, but there wouldn't have been much of a brain there. From that stage the evolution of nervous systems proceeds independently in many lines, including two that led to large brains of different design.

On our lineage, the chordate design emerges, with a cord of nerves down the middle of the animal's back and a brain at one end. This design is seen in fish, reptiles, birds, and mammals. On the other side, the cephalopods' side, a different body plan evolved, and a different kind of nervous system. These nervous systems are more *distributed*, less centralized, than ours. Invertebrates' neurons are often collected into many *ganglia*, little knots that are spread through the body and connected to each other. The ganglia can be arranged in pairs, linked by connectors that run along the body and across it, like lines of latitude and longitude. This is sometimes called a "ladder-like" nervous system, and it does look like a ladder embedded within the body. The ancestral cephalopods probably had nervous systems something like this, so when evolution multiplied their neurons, the multiplication took place on this design.

In that expansion, some ganglia became large and complex, and new ones were added. Neurons concentrated at the front of the animal, forming something more and more like a definite brain. The old ladder-like design was partly submerged, but only partly, and the underlying architecture of cephalopod nervous systems remains quite different from our own.

Perhaps most oddly, the esophagus, the tube that carries food from the mouth into the body, passes through the middle of the central brain. This seems all wrong; surely there was never supposed to be a brain *there*. If an octopus eats something sharp which pierces the side of its "throat," the sharp object goes straight into its brain. Octopuses have been discovered with exactly this problem.

Further, much of a cephalopod's nervous system is not found within the brain at all, but spread throughout the body. In an octopus, the majority of neurons are in the arms themselves—nearly twice as many as in the central brain. The arms have their own sensors and controllers. They have not only the sense of touch, but also the capacity to sense chemicals—to smell, or taste. Each sucker on an octopus's arm may have 10,000 neurons to handle taste and touch. Even an arm that has been surgically removed can perform various basic motions, like reaching and grasping.

How does an octopus's brain relate to its arms? Early work, looking at both behavior and anatomy, gave the impression that the arms enjoyed considerable independence. The channel of nerves that leads from each arm back to the central brain seemed pretty slim. Some behavioral studies gave the impression that octopuses did not even track where their own arms might be. As Roger Hanlon and John Messenger put it in their book *Cephalopod Behaviour*, the arms seemed "curiously divorced" from the brain, at least in the control of basic motions.

The internal coordination of each arm can be quite graceful, too. When an octopus pulls in a piece of food, the grasping by the very end of the arm creates two waves of muscle activation, one heading inward from the tip, and the other heading outward from the base. Where these two waves meet, a joint is formed that is something like a temporary elbow. The nervous systems in each arm also include loops in the neurons (*recurrent* connections, in the jargon) that may give the arm a simple form of short-term memory, though it's not known what this system does for the octopus.

Octopuses can pull themselves together in some contexts, though, especially when it matters. As we saw at the beginning of this chapter, when you encounter and approach an octopus in the wild and pause in front of it, in at least some species the octopus sends out *one* arm to inspect you. Often a second arm follows, but it's just one that comes out first, as the animal watches. This suggests a kind of deliberateness, an action guided by the brain. Below is a video frame from Octopolis that also suggests such a view. One octopus, in the center of the frame, leaps toward another on the right, a single arm cocked to seize its foe.

Some sort of mixture of localized and top-down control might be operating. The best experimental work I know that bears on this topic comes out of Binyamin Hochner's laboratory at the Hebrew University of Jerusalem. A 2011 paper by Tamar Gutnick, Ruth Byrne, and Michael Kuba, along with Hochner, described a very clever experiment. They asked whether an octopus could learn to guide a single arm along a maze-like path to a specific place in order to obtain food. The task was set up in such a way that the arm's own chemical sensors would not suffice to guide it to the food; the arm would have to leave the water at one point to reach the target location. But the maze walls were transparent, so the target location could be seen. The octopus would have to guide an arm through the maze with its eyes.

It took a long while for the octopuses to learn to do this, but in the end, nearly all of the octopuses that were tested succeeded. The eyes *can* guide the arms. At the same time, the paper also noted that when octopuses are doing well with this task, the arm that's finding the food appears to do its own local exploration as it goes, crawling and feeling around. So it seems that two forms of control are working in tandem: there is central control of the arm's overall path, via the eyes, combined with a fine-tuning of the search by the arm itself.

~ Body and Control

Half a billion neurons—why so many? What do they do for the animal? In the previous chapter I emphasized the expense of this machinery. Why did cephalopods go down this unusual evolutionary road? Nobody knows the answer to this, but I'll sketch some possibilities. The question arises to some degree for nearly all cephalopods, but I'll focus on octopuses.

Octopuses are predators, and they hunt by moving, rather than waiting in ambush. They rove around, often on reefs and shallow sea floors. When animal psychologists try to explain the evolution of a large brain, they often begin by looking at the social life of the animal. The complexities of social life seem to frequently give rise to high intelligence. Octopuses are not very social. In the final chapter I'll look at exceptions to this, but social life is not a big part of the octopus story. A factor that seems more important is all that roving and hunting. To sharpen this idea up I'll adapt some ideas developed in the 1980s by the primatologist Katherine Gibson. She was looking for an account of why some mammals evolved large brains, and didn't consider their application to anything like an octopus, but I think her ideas might be relevant here, too.

Gibson distinguished two different ways of foraging for food. One way is to specialize on a food that requires little manipulation and can be handled the same way in every case. Her example was a frog catching flying insects. She contrasted this with "extractive" foraging, the kind that involves adapting choices to circumstances, removing food from protective shells and casings, and doing so in a flexible and context-sensitive way. Compare the frog with a chimp, who wanders about searching for a variety of things to eat, many of which require manipulation and extraction once they're found—nuts, seeds, termites in their nests. Gibson's description of this flexible and demanding style of searching for food fits octopuses well. For many octopuses, crabs are at the top of the food preference list, but various additional animals, from scallops to fish (and other octopuses) also count as food, and dealing with shells and other defenses is often a significant task.

David Scheel, who works mostly with the giant Pacific octopus, feeds his animals whole clams, but as his local animals in Prince William Sound do not routinely eat clams, he has to teach them about the new food source. So he partly smashes a clam and gives it to the octopus. Later, when he gives the octopus an intact clam, the octopus knows that it's food, but does not know how to get at the meat. The octopus will try all sorts of methods, drilling the shell and chipping the edges with its beak, manipulating it in every way possible . . . and then eventually it learns that its sheer strength is sufficient: if it tries hard enough, it can simply pull the shell apart.

This style of hunting and foraging makes good sense of the exploratory, curious side of the octopus psyche, especially their engagement with novel objects. This factor is more applicable to octopuses than to cuttlefish and squid, which engage in less complicated manipulation of their food. Some cuttlefish have very large brains—perhaps even larger, as a fraction of the body, than octopuses. That is quite a mysterious fact at the moment, and less is known about what cuttlefish can do.

While octopuses are not very social, in the usual sense—the sense that involves spending a lot of time with other octopuses—their engagement with other animals as predators and as prey is "social" in a way. Those situations often require that an animal's actions be tuned to the actions and perspectives of others, including what those others can see and what they're likely to do. The demands of "social" life, in the within-species sense, have similarities to the demands of some kinds of hunting, and avoiding being hunted oneself.

Those features of the octopus lifestyle are probably part of the story behind its large nervous system. I now want to put another idea on the table as well. In chapter 2 I contrasted *sensory-motor* views and *action-shaping* views of the evolution of nervous systems. The action-shaping approach is less familiar, and it took some effort, historically, to develop it. The central idea is that rather than mediating between sensory input and behavioral output, the first nervous systems came to exist as solutions to a problem of pure coordination within the organism—the problem of how to coordinate the micro-acts of parts of the body into the macro-acts of the whole.

The cephalopod body, and especially the octopus body, is a unique object with respect to these demands. When part of the molluscan "foot" differentiated into a mass of tentacles, with no joints or shell, the result was a very unwieldy organ to control. The result was also an enormously *useful* thing, *if* it could be controlled. The octopus's loss of almost all hard parts compounded both the challenge and the opportunities. A vast range of movements became possible, but they had to be organized, had to be made coherent. Octopuses have not dealt with this challenge by imposing centralized governance on the body;

rather, they have fashioned a mixture of local and central control. One might say the octopus has turned each arm into an intermediate-scale actor. But it also imposes order, top-down, on the huge and complex system that is the octopus body.

The demands of pure coordination, which might have been important in the early evolution of nervous systems, also here take on a latter-day role. They may have been responsible for much of the multiplication of neurons in the octopus; those neurons are needed just to make the body controllable.

Though solving the problem of coordination would explain the nervous system's *size*, it would not explain the octopus's intelligent and flexible behavior. A well-coordinated animal could also be a rather uninventive animal. A more complete approach to the octopus might then combine these ideas about action-shaping with the ideas about foraging and hunting that I borrowed from Gibson earlier; those ideas would explain the animal's inventiveness, curiosity, and sensory acuity. Or the story might, more tendentiously, go like this. A large nervous system evolves to deal with coordination of the body, but the result is so much neural complexity that eventually other capacities arise as byproducts, or relatively easy additions to what the demands of action-shaping have built. I said "or" just above—byproducts *or* additions—but this is definitely an "and/or." Some capacities—such as recognition of individual people—might be by-products, while others—such as problem solving—are the results of the evolutionary modification of the brain in response to the octopus's opportunistic lifestyle.

In this picture, neurons first multiply because of the demands of the body, and then sometime later, an octopus wakes up with a brain that can do more. Certainly it seems that *some* of its impressive behavior is fortuitous, from an evolutionary point of view. Remember again those surprising behaviors in captivity, the mischief and craft, the engagement with humans. There is, it seems, a kind of mental surplus in the octopus.

~ *Convergence and Divergence*

I described how the early history of animals, insofar as we know it, led to a fork with one path running forward to chordates, like us, and the other leading to cephalopods, including the octopus. Let's take stock and compare what arose down the two evolutionary lines.

The most dramatic similarity is the eyes. Our common ancestor may have had a pair of eyespots, but it did not have eyes anything like ours. Vertebrates and cephalopods separately evolved "camera" eyes, with a lens that focuses an image on a retina. The capacity for learning of several kinds is also seen on both sides. Learning by attending to reward and punishment, by tracking what works and what does not work, seems to have been invented independently several times in evolution. If it was present in the human/octopus common ancestor, it was greatly elaborated down each of the two lines. There are also more subtle psychological similarities. Octopuses, like us, seem to have a distinction between short-term and long-term memory. They engage in play with novel objects that aren't food and have no apparent use. They seem to have something like sleep. Cuttlefish appear to have a form of *rapid eye movement* (REM) sleep, like the sleep in which we dream. (It's still unclear whether there's REM-like sleep in octopuses.)

Other similarities are more abstract, such as an involvement with individuals, including the ability to recognize particular humans. Our common ancestor surely could not do anything like this. (It's hard to imagine what that simple little creature would have taken its world to contain.) This ability makes sense if an animal is social or monogamous, but octopuses are not monogamous, have haphazard sex lives, and seem not very social. There's a lesson here about the ways that smart animals handle the stuff of their world. They carve it up into objects that can be re-identified despite ongoing changes in how those objects present themselves. I find this a striking feature of the octopus mind—striking in its familiarity, its similarity to our own.

Some features show a mixture of similarity and difference, convergence and divergence. We have hearts, and so do octopuses. But an octopus has *three* hearts, not one. Their hearts pump blood that is blue-green, using copper as the oxygen-carrying molecule instead of the iron which makes our blood red. Then, of course, there is the nervous system—large like ours, but built on a different design, with a different set of relationships between body and brain.

The octopus is sometimes said to be a good illustration of the importance of a theoretical movement in psychology known as *embodied cognition*. These ideas were not developed to apply to octopuses, but to animals in general, including ourselves, and this view has also been influenced by robotics. One central idea is that our body itself, rather than our brain, is responsible for some of the "smartness" with which we handle the world. Our body's own structure encodes some information about the environment and how we must deal with it, so not all this information needs to be stored in the brain. The joints and angles of our limbs, for example, make motions such as walking naturally arise. Knowing how to walk is partly a matter of having the right body. As Hillel Chiel and Randall Beer put it, an animal's body structure creates both *constraints and opportunities*, which guide its action.

Some octopus researchers have been influenced by this way of thinking, especially Benny Hochner. Hochner believes these ideas can help us grasp the octopus/human differences. Octopuses have a *different embodiment*, which has consequences for their different kind of psychology.

I agree with that last point. But the doctrines of the embodied cognition movement do not really fit well with the strangeness of the octopus's way of being. Defenders of embodied cognition often say that the body's shape and organization encodes information. But that requires that there *be* a shape to the body, and an octopus has less of a fixed shape than other animals. The same animal can stand tall on its arms, squeeze through a hole little bigger than its eye, become a streamlined missile, or fold itself to fit into a jar. When advocates of embodied cognition such as Chiel and Beer give examples of how bodies provide resources for intelligent action, they mention the distances between parts of a body (which aid perception) and the locations and angles of joints. The octopus body has none of those things—no fixed distances between parts, no joints, no natural angles. Further, the relevant contrast in the octopus case is not "body rather than brain"—the contrast usually emphasized in discussions of embodied cognition. In an octopus, the nervous system as a whole is a more relevant object than the brain: it's not clear where the brain itself begins and ends, and the nervous system runs all through the body. The octopus is suffused with nervousness; the body is not a *separate* thing that is controlled by the brain or nervous system.

The octopus, indeed, has a "different embodiment," but one so unusual that it does not fit any of the standard views in this area. The usual debate is between those who see the brain as an all-powerful CEO and those who emphasize the intelligence stored in the body itself. Both views rely on a distinction between brain-based and body-based knowledge. The octopus lives outside both the usual pictures. Its embodiment *prevents* it from doing the sorts of things that are usually emphasized in the embodied cognition theories. The octopus, in a sense, is *disembodied*. That word makes it sound immaterial, which is not, of course, what I have in mind. It has a body, and is a material object. But the body itself is protean, all possibility; it has none of the costs and gains of a constraining and action-guiding body. The octopus lives outside the usual body/brain divide.

4

FROM WHITE NOISE
TO CONSCIOUSNESS

What It's Like

What does it feel like to be an octopus? To be a jellyfish? Does it feel like anything at all? Which were the *first* animals whose lives felt like something to them?

At the start of the book I quoted William James's plea for "continuity" in our understanding of the mind. The elaborate forms of experience found in us derived from simpler forms in other organisms. Consciousness surely did not, James said, suddenly *irrupt* into the universe fully formed. The history of life is a history of intermediates, shadings-off, and gray areas. Much about the mind lends itself to a treatment in those terms. Perception, action, memory—all those things creep into existence from precursors and partial cases. Suppose someone asks: Do bacteria *really* perceive their environment? Do bees really *remember* what has happened? These are not questions that have good yes-or-no answers. There's a smooth transition from minimal kinds of sensitivity to the world to more elaborate kinds, and no reason to think in terms of sharp divides.

For memory, perception, and so on, this gradualist attitude makes a lot of sense. But the other side of the coin is subjective experience, the feel of our lives. Many years ago Thomas Nagel used the phrase *what it's like* in an attempt to point us toward the mystery posed by subjective experience. He asked: What is it like to be a bat? It's probably like *something*, but very different from what it's like to be a human. The term "like" is misleading here, as it suggests that the problem hinges on issues of comparison and similarity—*this* feeling is like *that* feeling. Similarity is not the issue. Rather, there *is a feel* to much of what goes on in human life. Waking up, watching the sky, eating—these things all have a feel to them. That's what has to be understood. But when we take an evolutionary and gradualist perspective, this takes us to strange places. How can the fact of life feeling like something slowly creep into being? How can an animal be halfway to having it feel like something to be that animal?

~ Evolution of Experience

I aim to make progress on those problems here. I don't claim to solve them entirely, but to take us closer to the goal James laid down. I'll set the topic up as follows. *Subjective experience* is the most basic phenomenon that needs explaining, the fact that life feels like something to us. People sometimes now refer to this as explaining *consciousness*; they take subjective experience and consciousness to be the same thing. Instead, I see consciousness as one form of subjective experience, not the only form. For an example that motivates this distinction, take the case of pain. I wonder whether squid feel pain, and whether lobsters and bees do. I take this question to mean: Does damage feel like anything to a squid? Does it feel *bad* to them? This question would often now be expressed by asking whether squid are conscious. That always sounds misleading to me, as if it's asking too much of the squid. To use an older term, if it feels like something to be a squid or octopus, then these are *sentient* beings. Sentience comes before consciousness. Where does sentience come from?

It's not a soul-like substance that is somehow added to the physical world, as *dualists* think. Nor is it something that pervades all of nature, as *panpsychists* believe. Sentience is brought into being somehow from the evolution of sensing and acting; it involves being a living system with a point of view on the world around it. If we take that approach, though, a perplexity we run into immediately is the fact that those capacities are so widespread—they are found far outside the organisms that are usually thought to have experience of some kind. Even bacteria sense the world and act, as we saw in chapter 2. A case can be made that responses to stimuli, and the controlled flow of chemicals across boundaries, are an elementary part of life itself. Unless we conclude that all living things have a modicum of subjective experience—a view I don't regard as insane, but surely one that would need a lot of defense—there must be something about the way *animals* deal with the world that makes a crucial difference.

One way to approach this question is just to talk about the complexity of different kinds of organisms, and the complexity of their dealings with the world. But there are many kinds of complexity, and we want something more specific. I'll look now at one such factor—something that I'm sure is part of the story, though it is not easy to see exactly where it fits in. In animal evolution, along with the sheer elaboration of sensing and acting, there's the evolution of new kinds of connection between these activities, especially connections that loop, that involve feedback.

For an organism like you or me, here are some familiar facts. What you'll do next is affected by what you're now sensing; and also, what you'll *sense* next is affected by what you now *do*. You read, and turn the page, and turning the page will affect what you see. Sensing and acting each affect the other. We know this explicitly and can talk about it, but their intertwining also affects in more fundamental ways how things feel, in quite a raw sense of "feel."

Consider the case of *tactile vision substitution systems* (TVSS), a technology for the blind. A video camera is attached to a pad that sits on the blind person's skin (for example, on their back). Optical images picked up by the camera are transformed into a form of energy (vibrations, or electrical stimulation) that can be felt on the skin. After some training with this device, the wearers start to report that the camera gives them an experience of *objects located in space*, not just a pattern of touches on their skin. If you are wearing such a system and a dog walks past, for example, the video system will make a moving pattern of presses or vibrations on your back, but under some circumstances this will not be experienced as vibrations on your back; instead you'll experience an object moving out in front of you. This happens, though, only when the wearer is able to *control the camera*, to act and influence the incoming stream of stimulation. The user of the device has to be able to move the camera closer, change its angle, and so on. The simple way to do this is to attach the camera to the person's body. Then the wearer can make objects loom, and come and go from the visual field. Subjective experience is intimately tied here to the interaction between behavior and sensory input. The moment-to-moment feedback between sensing and acting affects how sensory input itself feels.

Though the idea that our actions affect what we perceive seems routine and familiar, philosophers through many centuries did not treat it as especially important. In philosophy, this is the territory of unorthodoxies, of works beside, rather than within, the main development of ideas. That is true even in recent years. Instead, a huge amount of work has looked at a small *piece* of the total picture; it has looked at the link between what comes in through the senses and the thoughts or beliefs that result. Little was usually said about the link to action, and even less about the way action affects what you sense next.

Some philosophers have always disliked this obsession with sensory input, with receptivity, seen in theories of the mind. But their response was to reject the importance of input in a wholesale way, and to try to tell a story about the self-determining organism, about the subject as *source*, imposing itself on the world. This is a kind of overcompensation, as if philosophers are only able to concentrate on one side at a time. It is a big thing, apparently not easily achieved, to accept that there is *traffic*, a to-and-fro.

In everyday experience there are two causal arcs. There is a sensory-motor arc, linking our senses to our actions, and a *motor-to-sensory* arc as well. Why turn the page? Because doing so will influence what you'll see next. The second arc is not as tightly controlled as the first, because it extends into external, public space, rather than remaining inside the skin. Perhaps as you turn the page, someone grabs the book, or grabs you. The sense-to-motor and the motor-to-sense pathways are not on a par. But the neglected junior partner, the effect of action on what we sense next, is surely important. This, after all, is *why* we do much of what we do: to control what our senses will encounter.

Philosophers often use the metaphor of a *stream* of experience. Experience, they say, is something like a river in which we are immersed. This image is quite misleading, though, as the flow in a river is almost entirely outside our control. We might change our location within the river, swimming to one spot or another, and that gives us some control over what we'll encounter. But in real life we can usually do much more than that; we can re-shape the things themselves that we interact with. Rivers are rather resistant to such efforts when we're alone in the middle of them.

What you sense next has two sources: what you just did, and what the larger world beyond you is up to. The overall shape of the cause-effect relations looks like this.

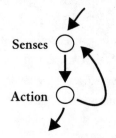

Two arrows lead into the senses. They have different roles in different contexts and sometimes one is more important than the other, but they are both almost always there.

Loops that link actions back to the senses are not only seen in us. They are present also in very simple forms of life. But they become more marked in animals, especially because animals can *do* more. The evolution of muscle, derived from tiny fiber-like elements inside cells, created a new means by which life impresses itself on the world. All living things affect their environment by making and transforming chemicals, and also by growing and sometimes by moving, but it is muscle that gives rise to rapid, coherent action on large spatial scales. It makes possible the *manipulation* of objects, the deliberate and rapid transformation of what is around us.

The evolution of animals is affected by these looping causal paths in a number of ways. Often these loops lead to a *problem*, as an animal attempts to work out what's going on around it. Some fish, for example, send out electric pulses for communication with other fish, and also electrically sense other things going on around them. The self-produced pulses will affect their own senses, though, and it may be difficult for a fish to distinguish the pulses it has made from electrical disturbances that are due to external things. To deal with this problem, whenever a fish emits a pulse it also sends a *copy* of the command around to the sensing system, enabling that system to counteract the effect of the pulse it has produced. The fish is tracking and registering the distinction between "self" and "other," between the effects on its senses of its own actions and the effects of events going on around it.

An animal need not send out electrical pulses to encounter this problem. As the Swedish neuroscientist Björn Merker notes, it results just from being able to move. An earthworm withdraws when something touches it—the touch might be a threat. But every time the worm crawls forward, it causes part of its body to be touched in just the same way. If it withdrew at every touch, it could never move at all. The worm succeeds in moving forward by canceling the effects of those self-produced touches.

For all organisms there is a distinction between self and the external world, even if only onlookers can see it. All organisms also affect the world outside them, whether they register that fact or not. Many animals, though, acquire their own glimpse, their own registration, of these facts, because action would be so difficult otherwise. Plants, in contrast, have quite rich senses but don't move. Bacteria move, but their simple senses don't threaten to tangle them, in the way seen in Merker's worm.

This interaction between perception and action is also seen in what psychologists call *perceptual constancies*. For us, an object can remain recognizable as the same object while our viewpoint on it changes. As you move closer to or further from a chair, it does not usually seem to grow, shrink, or move, because you tacitly compensate for the changes in appearance that are due to your actions, along with some changes that are not due to you, such as shifts in lighting conditions, and so on. Perceptual constancies are seen in a fairly wide range of animals, including octopuses and some spiders, as well as vertebrates. This ability has probably evolved independently in several different groups.

Another path in the evolution of experience leads to *integration*. As streams of information come in from different senses, they are brought together into a single picture. This is vivid in our own case; we experience the world in a way that ties together what we see with what we hear and touch. Our experience is usually of a unified scene.

This might seen inevitable, a consequence of having eyes and ears attached to the same brain, but it is not. It's one way of being wired up, and some animals do not integrate their experience nearly as much as we do. For example, in many animals the eyes are on each side of the head, not the front. The eyes then have separate visual fields, either largely or entirely, and each connects just to one side of the brain. In such an animal it is easy for scientists to control what each side is exposed to, by masking one eye. Then we can ask a question that might seem to have an obvious answer: If we show something to only one side of the brain, does the other side get the information too? We are not looking at damaged or altered animals here, so the two sides of the brain have all their natural connections. One would think that the information should get across. Why should evolution set things up so that only half the animal knows what has been seen? But when this question was studied in pigeons, it turned out that the information was *not* being passed across. The pigeons were trained to do a simple task with one eye masked, then each pigeon was tested on the same task while being forced to use the other eye. In a study using nine birds, eight of them did not show any "inter-ocular transfer" at all. What seemed to be a skill learned by the whole bird was in fact available to only half the bird; the other half had no idea.

These experiments have also been done on octopuses. An octopus trained on a visual task using just one eye initially remembered the task only when tested with the same eye. With extended training, they could perform the task using the other eye. The octopuses were unlike the pigeons in that some information did get across; they were unlike us in that it did not get across easily. In more recent years, animal researchers such as Giorgio Vallortigara of the University of Trieste have uncovered a number of similar "fissures" in information processing related to the separation between the two halves of the brain. A variety of species appear to be more reactive to predators seen on the left side of their visual field. Several kinds of fish, and even tadpoles, seem to prefer to position themselves so that an image of another individual of the same species is on their left side. On the other hand, various animals are better at perceiving what's on their right side when the aim is to detect food.

This specialization seems to have clear disadvantages, leaving the animal vulnerable to attack on one side, or less able to find food on one side. Vallortigara and others think it might make good sense, though. If different tasks require different kinds of processing, it may be best to have a brain with specialized sides that deal with each task, and not tie them too closely together.

These findings are reminiscent of experiments on "split brain" humans. In cases of severe epilepsy, it sometimes helps the patient to cut the *corpus callosum*, the connector between the left and right hemispheres that make up the upper part of the human

brain. People tend to behave fairly normally after these opera-
tions, and it took a while for researchers to realize that anything unusual was going on. But if the different halves of such a
patient's brain are exposed to different stimuli, quite dramatic
disunity often emerges. The operation seems to have given rise
to two intelligent selves, with different experiences and skills, in
a single skull. The left side of the brain usually controls language
(though not always), and when you talk to a split-brain patient, it
is the left side who speaks back. Though the right side cannot
usually speak, it can control the left hand. So it can choose objects
by touch, and draw pictures. In various experiments, different
images are provided to each side of the brain. If the person is asked
what they have seen, their verbal response will follow what was
shown to the left side of the brain, but the right side—controlling
the left hand—may disagree. The special kind of mental frag-
mentation seen in split-brain humans seems to be a routine part of
many animals' life.

Animals seem to have a range of ways of dealing with this
situation. In the case of birds, incoming visual information can
be even more fragmented than it was in those eye-masking experiments I described above. In birds like pigeons, each retina
has two different "fields," the yellow field and the red field. The
red field sees a small area in front of the bird where there is
binocular vision, and the yellow field sees a larger area that the
other eye cannot access. Pigeons not only failed to transfer information between eyes; they also did quite badly at transfer between
different regions of the *same* eye. This might explain some distinctive bird behaviors. Marian Dawkins did a simple experiment
showing a novel object (a red toy hammer) to hens, who were allowed to approach and inspect it. She found that hens approached
such an object in a weaving way that seemed designed to give
the different parts of each eye access to it. That, apparently, is

the way the whole brain gets access to the object. The weaving gaze of a bird is a technique designed to slosh the incoming information around.

To some degree, unity is inevitable in a living agent: an animal is a whole, a physical object keeping itself alive. But in other ways, unity is optional, an achievement, an invention. Bringing experience together—even the deliverances of the two eyes—is something that evolution may or may not do.

~ Latecomer versus Transformation

The story I am working toward is one of gradual change: as sensing, acting, and remembering became more elaborate, the feel of experience became more complex along the way. Our own case shows us that subjective experience is not an all-or-nothing matter. We know half-conscious states of various kinds, such as waking from sleep. Evolution includes an awakening on a different time scale.

But perhaps all that is a mistake. A gradual development of subjectivity from simple and early forms is one option, but perhaps the best evidence we have tells against it, evidence that comes from our own brains.

One path to this view begins with an accident, a case of carbon monoxide poisoning from a shower's bad water heater in 1988, which led to a case of brain damage in a woman known only as "DF." As a result of the accident, DF felt almost blind. She lost all experience of the shapes and layout of objects in her visual field. Only vague patches of color remained. Despite this, it turned out that she could still *act* quite effectively toward the objects in space around her. For example, she could post letters through a slot that was placed at various different angles. But she could not describe the angle of the slot, or indicate it by pointing. As far as subjective experience goes, she couldn't see the slot at all, but the letter reliably went in.

DF has been studied extensively by the vision scientists David Milner and Melvyn Goodale. By linking her case to other kinds of brain damage and to earlier work in anatomy, Milner and Goodale put together a theory of what's going on—in us, as well as in special cases like DF's. They argue that there are two "streams" by which visual information moves through the brain. The *ventral stream*, which takes a lower path through the brain, is concerned with categorization, recognition, and description of objects. The *dorsal stream*, which runs above it, closer to the top of the head, is concerned with real-time navigation through space (avoiding obstacles as you walk, getting the letter through the slot). Milner and Goodale argue that our subjective experience of vision, the feel of the visual world, comes only from the ventral stream. The dorsal stream does its work unconsciously, both in DF and in ourselves. After her accident, DF lost her ventral stream, and hence felt almost blind—even as she walked around the obstacles before her.

A simple interpretation of these cases holds that you need the ventral stream to have any experience of what's coming in through your eyes at all. That is probably too simple. It's likely that dorsal stream vision feels like something, though it doesn't feel much like seeing. The details of these two "streams" are less important than the larger surprise coming out of this work. That is the fact that quite complicated processing of visual information—processing that runs all the way from eyes, through brain, to legs or hands—can take place without the subject experiencing any of this *as seeing*. Milner and Goodale link this discovery to what I described a bit earlier as the integration of sensory information. They think that the activity in our brains that leads to visual experience is the building of a coherent "inner model" of the world. It's certainly reasonable to think that building an integrated model of this kind has effects on subjective experience. But perhaps without such a model there is no subjective experience at all?

Milner and Goodale discuss various animals whose perception of the world is less integrated than ours. In the 1960s, David Ingle rewired the nervous systems of some frogs by means of surgery (he was aided by the fact that frog nervous systems regenerate unusually well). By crossing some wires in the brain, he was able to produce a frog that snapped at prey to its left when the prey was really to its right, and vice versa. It saw prey in a left-right reversed way. But this rewiring of part of the visual system did not affect all of the frog's visual behavior. The frogs behaved normally when they were using vision to get around a barrier. They behaved as if some parts of the visual world were reversed, and other parts were normal. Here is Milner and Goodale's comment:

> So what did these rewired frogs "see"? There is no sensible answer to this. The question only makes sense if you believe that the brain has a single visual representation of the outside world that governs all of an animal's behavior. Ingle's experiments reveal that this cannot possibly be true.

Once you accept that a frog does not have a unified representation of the world, and instead has a number of separate streams that handle different kinds of sensing, there is no need to ask what the frog sees: in Milner and Goodale's words, "the puzzle disappears."

Perhaps one puzzle disappears, but another is raised. What does it feel like to be a frog perceiving the world in this situation?

I think that Milner and Goodale are suggesting that it feels like *nothing*. There is no experience here because the machinery of vision in frogs is not doing the sorts of things it does in us that give rise to subjective experience.

Milner and Goodale's comments illustrate, in one form, an idea that quite a few people who work in this area now accept. The senses can do their basic work, and actions can be produced, with all this happening "in silence" as far as the organism's experience is concerned. Then, at some stage in evolution, extra capacities appear that do give rise to subjective experience: the sensory streams are brought together, an "internal model" of the world arises, and there's a recognition of time and self.

What we experience, in this view, *is* the internal model of the world that complex activities in us produce and sustain. Feeling starts *there*—or, at least, it begins to creep into existence when these capacities creep into existence—in the brains of monkeys and apes, dolphins, perhaps other mammals, and some birds. When we think of simpler animals as having subjective experience, according to this view, we're projecting onto them a fainter version of our *own* kind of experience. This is a mistake because our experience relies on features they just don't possess.

A view like this has been defended also by the neuroscientist Stanislas Dehaene, whose laboratory near Paris has done some of the most penetrating work on this topic over the last twenty years or so. Dehaene and his coworkers have spent years looking at perception right at the edge of consciousness—images that come and go just a bit too quickly for subjects to know they are seeing them, or that are presented while attention is diverted, and that nevertheless affect what the subject thinks and does. It turns out that we often process this unexperienced information in quite sophisticated ways. For example, sequences of words can be flashed so quickly that a person has no idea they were shown at all. But sequences with incongruous meanings—such as "very happy war"—are registered by the brain differently from combinations with more reasonable meanings—"not happy war." One might think that conscious thought is necessary to distinguish such meanings, but this is not so.

We can do a lot without consciousness, Dehaene thinks, but some things we can't do. We can't unconsciously perform a task that is novel, rather than routine, and requires a series of acts, step-by-step. We can unconsciously learn associations between experiences—learn to expect A when you see B—but only if B and A come close together. Once there is a significant gap between them, we can learn the association only if we are aware of it. You can learn to blink when you see a light if the light is followed by an irritating puff of air, but only if the light and puff are very close together. Once the light and puff are separated by a second or so, the association can't be learned unconsciously. What we've learned over the last thirty years or so, Dehaene thinks, is that there's a particular *style* of processing—one that we use to deal especially with *time*, *sequences*, and *novelty*—that brings with it conscious awareness, while a lot of other quite complex activities do not.

Back in the 1980s, in one of the first modern attempts to explain consciousness, the neuroscientist Bernard Baars introduced the *global workspace* theory. Baars suggested that we are conscious of the information that has been brought into a centralized "workspace" in the brain. Dehaene adopted and developed this view. A related family of theories claim that we are conscious of whatever information is being fed into *working memory*, a special kind of memory which holds an immediate store of images, words, and sounds that we can reason with and bring to bear on problems. My colleague at the City University of New York, Jesse Prinz, has defended a view of this kind. If you think that a global workspace is needed for subjective experience, or a special kind of memory, or some other mechanism along these lines, you'll hold that only complex brains that are fairly similar to ours can give rise to experiences that feel like something. These brains will probably be found outside of people, but perhaps only in mammals and birds. The result is what I'll call *latecomer* views about subjective experience. These views don't hold that the lights went on in a sudden flash, but they do hold that the "waking up" came late in the history of life and was due to features that are clearly seen only in animals like us.

When I described a number of these theories just above, including the theories of Baars, Dehaene, and Prinz, I said they were theories of *consciousness*. I used that word because that's the word they use. It's sometimes hard to work out how these theories relate to my own target here: subjective experience in a very broad sense. I treat subjective experience as a broad category and consciousness as a narrower category within it—not everything that an animal might *feel* has to be *conscious*. A person might then say that a "global workspace" is necessary for consciousness without it being needed for the most basic kind of subjective experience. Not only is this possible, but I think it's approximately right. In the literature I'm describing here, it's often hard to work out what people think about this. But some of these people think there's no distinction between consciousness and subjective experience; they say that they're giving us a theory of when a mental activity *feels like something*.

The work that inspires latecomer views of experience has led to a great deal of progress. People like Dehaene have found a way *in* to the study of human consciousness, a path of a kind that seemed a fantasy not many years ago. We should not hang on to an alternative picture just because it's more generous or feels right. But I do think that arguments can be given against the latecomer view, and certainly there's an alternative to consider. I'll call this the *transformation* view. It holds that a form of subjective experience preceded late-arising things like working memory, workspaces, the integration of the senses, and so on. These complexities, when they came along, transformed what it feels like to be an animal. Experience has been reshaped by these features, but it was not brought into being by them.

The best argument I can offer for this alternative view is based on the role in our lives of what seem like old forms of subjective experience that appear as *intrusions* into more organized and complex mental processes. Consider the intrusion of sudden pain, or of what the physiologist Derek Denton calls the *primordial emotions*—feelings which register important bodily states and deficiencies, such as thirst or the feeling of not having enough air. As Denton says, these feelings have an "imperious" role when they are present: they press themselves into experience and can't easily be ignored. Do you think that those things (pain, shortness of breath, etc.) *only feel like something* because of sophisticated cognitive processing in mammals that has arisen late in evolution? I doubt it. Instead, it seems plausible that an animal might feel pain or thirst without having an "inner model" of the world, or sophisticated forms of memory.

Let's look at the case of pain. One might initially say it's obvious that even simple animals respond to pain in a way that indicates they feel it, squirming and wriggling in distress. But things are not so straightforward. Many responses to bodily damage that seem to involve pain probably do not. For example, rats with a severed spinal cord, and hence no channel from the site of body damage to the brain, can exhibit some of what looks like "pain behavior," and can even show a form of learning that responds to the damage. Various reflex responses in animals might look to us like pain, because we empathize with them. We need to go past these mere appearances.

Fortunately, we can. The most telling evidence is based on pain-related behaviors that are too flexible to be dismissed as reflexes, even though the animals in question have quite different brains from ours, and are not likely to meet the requirements of the "latecomer" views. Here is an example from fish. Zebrafish were tested first to see which of two environments they preferred. They were then injected with a chemical suspected to cause pain, and in some cases, the less preferred environment had a painkiller dissolved in it. The fish now preferred this environment, but only when it contained dissolved painkiller. They made a choice they'd not normally make, and they made it in a situation where the idea of a more painful or less painful *environment* would be quite novel to them: evolution could not have set them up with a reflexive reaction to this situation.

Similarly, in a study in chickens, birds with damaged legs chose a food that would usually be less preferred, provided that it contained painkillers. Robert Elwood has done experiments of a similar kind in hermit crabs, the small crabs who live in shells made by various mollusks. Hermit crabs are arthropods, relatives of insects. Elwood gave the crabs small electric shocks, and found they could be induced to leave their shell by a shock. But not always: they were more reluctant to leave a higher-quality shell than a low-quality one—they had to be shocked more. They were also more likely to put up with the shock when the odor of a predator was around and the shell was more valuable for protection.

Tests of this kind don't suggest that *all* animals feel pain. Insects are in the same large animal group (arthropods) as crabs. Insects appear to behave normally, to the extent that they physically can, even after quite severe injuries. They don't groom or protect injured parts of their body, but keep doing whatever they were doing. Crabs and some shrimp, in contrast, will groom injured areas. You can still doubt that these animals feel anything, yes. But you can doubt that about your next-door neighbor. Skepticism is always possible, but a case is being built here. These results do provide support for a view of pain as a basic and widespread form of subjective experience, one present in animals with very different brains from ours.

In this picture, there are early and simple forms of subjective experience that are then transformed as evolution makes nervous systems more complicated. With this transformation, new capacities—such as sophisticated kinds of memory—are added which have a subjective side, while other things that once contributed to experience might be pushed into the background. How might we imagine the earlier forms? This is perhaps impossible, as our imaginations are tied to our present-day, complicated minds. But let's try.

The title of this chapter borrows a phrase from a paper by Simona Ginsburg and Eva Jablonka. Two Israeli scientists working in different fields within biology, they wrote a paper a while ago trying to sketch the evolutionary origins of subjective experience. At one point in their paper, they offer a stab at a description of experience in a simpler and distant animal: *white noise*. Imagine a poorly differentiated buzz as the beginning of it all.

I keep coming back to that metaphor when I'm trying to get my mind around this topic. It *is* a metaphor—very much so. It's a metaphor of sound applied to organisms that, at least in most cases, probably could not hear at all. I'm not sure why the image stays so consistently with me. Somehow it seems to point in the right direction, with its evocation of a crackle of metabolic electricity, and the *shape* of the story suggested. That shape is one in which experience starts in an inchoate buzz, and becomes more organized.

In our own case, looking inside, we find that subjective experience has a close association with perception and control—with using what we sense to work out what to do. Why should this be? Why shouldn't subjective experience be associated with other things? Why isn't it brimful of basic bodily rhythms, the division of cells, life itself? Some people might say it *is* full of those things—more than we realize, anyway. I don't think so, and suspect there's a clue here. Subjective experience does not arise from the mere running of the system, but from the modulation of its state, from registering things that matter. These need not be external events; they might arise internally. But they are tracked because they matter and require a response. Sentience has some *point* to it. It's not just a bathing in living activity.

Ginsberg and Jablonka imagined their "white noise" as the first form of subjective experience. Perhaps, though, white noise corresponds to the *absence* of experience; it's what was present before subjective experience arose. Maybe that distinction takes the metaphor too far. Anyway, from some such state, the older forms of subjective experience arose—forms linked to the primordial emotions, pain and pleasure, feelings that must be acted on.

If this is right, some tentative conclusions might be drawn about the first animals with nervous systems, discussed in chapter 2. Suppose it's true that much of the work done by very early nervous systems was just pulling the animal together and making coordinated action possible. The patterned contraction of a swimming medusa is a present-day illustration, and the self-possessed lives that might have been lived by Ediacaran animals are also in this category. Here a nervous system acts mostly to generate and maintain an activity, and modulation of that activity is a more minor player. If so, perhaps this is a form of animal life that feels like nothing at all. The Cambrian—with all its richer forms of engagement with the world—would then be where simple experience begins.

This beginning would not have been a single event, or even a single extended process occurring on one evolutionary path. Instead, there would have been several such processes, taking place in parallel. By the time of the Cambrian, many of the different kinds of animals I've been discussing in this chapter had already branched off from one another—the branchings probably occurred in the Ediacaran, when all was quieter. By the Cambrian, the vertebrates were already on their own path (or their own collection of paths), while arthropods and mollusks were on others. Suppose it's right that crabs, octopuses, and cats all have subjective experience of some kind. Then there were at least three separate origins for this trait, and perhaps many more than three.

Later, as the machinery described by Dehaene, Baars, Milner, and Goodale comes on line, an integrated perspective on the world arises and a more definite sense of self. We then reach something closer to *consciousness*. I don't see that as a single definite step. Instead, I see "consciousness" as a mixed-up and overused but useful term for forms of subjective experience that are unified and coherent in various ways. Here, too, it is likely that experience of this kind arose several times on different evolutionary paths: from white noise, through old and simple forms of experience, to consciousness.

~ *The Case of the Octopus*

Let's now return to the octopus, our unusual and historically important animal. How does it fit in? What might *its* experience be like?

An octopus is, first, an organism with a large nervous system and a complex active body. It has rich sensory capacities and extraordinary capacities for behavior. If there is a form of subjective experience that comes along with sensing and acting in a living system, an octopus has plenty of that. But that's not all. In elusive and alien form, the octopus has some of the sophistications, some of the steps beyond the basics, described in this chapter.

Octopuses, of at least some species, have an opportunistic, exploratory style of interaction with the world. They are curious, embracing novelty, protean in behavior as well as in body. These features are reminiscent of what Stanislas Dehaene associates with consciousness in human mental life. As he says, the demands of novelty jolt us from unconscious routine into conscious reflection. An octopus's explorations are sometimes mixed with caution and sometimes with a puzzling recklessness. I noted in the previous chapter how my collaborator Matt Lawrence, diving near the Octopolis site, came across an octopus who seized his hand and led him over the sea floor, pulling him along behind it. We have no idea why it might have done this. In contrast, once while scuba diving at another site, I was hovering off the sea floor, anchored with a few fingers of one hand, as I photographed tiny sea slugs. I became aware of something below, and saw that a single slender octopus arm was slowly extending toward my fingers on the bottom, from a clump of seaweed next to me. The octopus was curled up in a ball in the weed, with most of its body hidden but one eye visible through a hole, sending one arm out cautiously as it watched. This was an act of exploration, accompanied by what seemed to be very close attention, keeping me in view as the arm went out. I was a novel object of uncertain importance. The seaweed provided both cover and a viewing hole. From this shelter, one arm was sent out to inspect, perhaps to taste.

Earlier I discussed *perceptual constancies*. These are abilities an animal has to re-identify objects despite changes in viewing conditions—distance, lighting, and so on. The animal must factor out the contribution of its own location and perspective to identify the object itself. Psychologists and philosophers often associate this ability with sophisticated, as opposed to rudimentary, forms of perception. Perceptual constancies show that an animal is perceiving external objects *as* external objects—as objects that can stay the same while the animal's vantage point changes. In an old 1956 experiment some octopuses were taught to approach particular shapes and avoid others. In some of the experiments, the relevant difference was between small and large square shapes. The octopus sat in a tank, a square was introduced at the other end, and the octopus had to approach some squares (for reward) and not approach others (or be punished, with electric shock). That was the routine, and the octopuses were able to do this. The researchers then say, almost in passing, that on "several" occasions, the octopus was presented with the small square at half the usual distance from its body. The small square would then initially look larger—or at least, the size of the squarish shape on its retina would be larger. In every such trial, the experimenters say, the octopus performed the right action for the square's real size. The octopus was able to factor out the change in distance.

A surprising thing about this report is that it is quite an important observation, and yet it's little more than a brief aside in the paper. No numbers are given for the cases that tested for perceptual constancy, and no one seems to have followed the idea up. If the finding is accepted, it shows that octopuses do have at least some perceptual constancies. So, apparently, do some other invertebrates, honeybees and some spiders; this is not one of the octopus's unique invertebrate achievements.

Octopuses are also good at navigation. Whenever I see an octopus wander from its den, I follow it if I can, and I've been taken on a lot of tours. If I don't get too close as they roam and explore, the octopuses often pay me no attention at all. The octopuses are usually foraging for food, and this takes them on long, rambling paths that eventually return to their dens. I am often surprised at how well they do this, as the ramblings can take a good fifteen minutes or so, traveling through quite murky water. If they head off from their den in one direction, they may well return to it from another. The tour has the form of a loop, not an out-and-back path. Some years ago, Jennifer Mather did a careful study of this kind of behavior, watching an octopus in the Caribbean as it went on hunting trips, and she charted looping paths of this kind. It's not known how octopuses do this—what sorts of landmarks, guides, and memories they make use of. But some octopus species are certainly good navigators.

Remember again that our most recent common ancestor—a worm-like creature in the Ediacaran—almost certainly had none of these skills. It seems that once an animal starts to lead an active and mobile life, full of controlled, goal-directed, and rapid movement, there are ways of seeing and handling the world that make more sense than others. Different animals have independently evolved perceptual constancies. Though in some ways they must see the world very differently from us, octopuses seem to deal with the world by identifying and re-identifying objects, and to have some grip on a distinction between self and other. When you are around an octopus, it's impossible not to think they can also direct considerable *attention* on objects, especially new ones.

In the previous section I discussed some work on pain behavior in fish, chickens, and crabs. Trying to work out how octopuses relate to pain is not easy. At Octopolis, our site in Australia, we once got a lot of video of a large male octopus engaged in a series of aggressive interactions, roaming and wrestling with others at the site. He often "stood tall," up on stretched legs, and sometimes raised his rear end high in the water, above his head. We think he was doing this to make himself look as large as possible; these displays were often precursors to an attack on another octopus. Once when he had his body positioned in this way, a small but vicious fish (a leatherjacket) darted in and bit him right on the rear. Here's the bite as it happened, with the fish at top center:

The octopus responded in a way that was very human-like, with a startled jump, arms going everywhere.

He then went straight back to beating up other octopuses. The bite was fortunate for us, as it left a noticeable mark, which we could use to identify this individual from some distance for the rest of that trip.

As we've seen, some animals tend and protect a wounded spot on their body. Our octopus bitten on the rear did not do that. His initial response suggested that he certainly felt the bite, but the aftereffects were not noticeable. We suspect this was because it was a minor injury and he was busy with ongoing pugilistic activities. A recent paper written by Jean Alupay and her colleagues looked carefully at pain-related behaviors, including wound tending, in another species of octopus. There was some reason to expect oddities, because some octopuses, including Alupay's species, nip off their own arms, when necessary, to escape predators. The study found that octopuses who'd had their arms crushed (not *too* crushed) in an experiment amputated them in some cases, not in all, and they all tended and guarded the injured site for some time. That tending and guarding is, as I discussed, usually seen as an indicator of pain.

In the case of the octopus, everything about experience is made more complicated by the unusual relations between brain and body. Let's assume the octopus has a kind of mixed control over what its arms do, an interpretation supported by the behavioral experiments discussed in chapter 3. Octopuses, as they evolved their complex behavioral abilities, opted for a partial delegation of autonomy to their arms. As a result, those arms are brimming with neurons and seem able to control some actions locally. Given that, what might octopus experience be like?

The octopus may be in a sort of hybrid situation. For an octopus, its arms are partly *self*—they can be directed and used to manipulate things. But from the central brain's perspective, they are partly *non-self* too, partly agents of their own.

Let's consider some analogies with our case, beginning with acts like blinking and breathing. These are activities that normally happen involuntarily, but through attention you can assert control over them. An octopus's arm movements have something like this combination. The analogy is imperfect, as breathing, while normally involuntary, can be subjected to very fine-grained control when you do intervene to breathe voluntarily. Attention is used to take over what is normally an automatic process. In the octopus, if the mixed-control interpretation is right, central guidance of the movements is never complete, and the peripheral system always has its say. To put it too anthropomorphically: you would send an arm out deliberately and *hope* the local fine-tuning goes right.

Action by an octopus, then, would mix elements that are usually distinct, or at least seem that way, in animals like us. When we act, the border between self and environment is usually fairly clear. If you move your arm, for example, you control the arm both on its general path and also in many fine details of its motions. Various other objects in the environment are not under your direct control at all, though they can be moved indirectly by manipulating them with your limbs. Uncontrolled movements by an object around you are usually a sign that it is not part of you (with partial exceptions for knee-jerk reflexes and the like). If you were an octopus, these distinctions would be blurred. To some extent you would guide your arms, and to some extent you would just watch them go.

To tell the story this way is to tell it from the vantage point of the "central octopus." That might be an error. In addition, I might be assuming too simple a contrast with the human case. When a person becomes able to play a musical instrument well, various actions—including adjustments—become too rapid to be controlled consciously. Bence Nanay, a philosopher based in Antwerp, also sent me a quite different interpretation of the octopus/human comparison. Bence thinks that some relationships that look weird and novel in the octopus case are present in our case too, if we look hard enough. They are usually invisible to us, but they are there. Suppose you are reaching for an object with your hand. If the location or size of the target you are reaching for suddenly changes, your reaching movement changes extremely quickly—in less than a tenth of a second. This is so fast that it is unconscious. Subjects in experiments don't notice the change—they don't notice that *they've* changed their own movement, and don't notice the change in the target object. When I say "subjects" in the experiment, I mean that if you ask the person if there was a change, they will say no. The person does not notice the change, but their arm alters its path.

As in the octopus, there's a top-down decision to extend the hand, but also a fine-tuning which is fast and unconscious. In the octopus case, the fine-tuning is greater—it's more than just fine-tuning—and it does not only happen quickly. The octopus might watch some of the arm's wandering as if it is a spectator. In us, these adjustments are too fast to see.

In the case of humans, these rapid adjustments of the arm

come from the brain, and they are visually guided. In the octopus, the motions are guided by the arm's own chemical and tactile senses, not by vision (though I'll qualify that statement in the next chapter; the issue is not quite so clear). In any case, Nanay's interpretation is that the octopus displays an extreme case of something that is also present in human action, though in a milder, less noticeable form. In the human case, there's a top-down command and then the addition of whatever fine-tuning is needed. In the octopus case, there's probably an ongoing interaction between commands from the center and decisions at the periphery. The arm is sent out, it wanders, and the octopus might respond by adjusting—perhaps by using attention, exerting some octopus willpower—to redirect the arm and keep it on track.

In the paper on "embodied cognition" I quoted earlier, Hillel Chiel and Randy Beer contrast an old and a new view of how action works. The old view has it that the nervous system is the "conductor of the body, choosing the program for the players and directing exactly how they play." Instead of this, they say, "the nervous system is one of a group of players engaged in jazz improvisation, and the final result emerges from the continued give and take between them." I'm not convinced by that as a general claim; I think it understates the role of the nervous system in most animals, to see it as one player among many. But in the case of the octopus, such a metaphor may well apply. The contrast now is not between the nervous system and the body, but between the central brain and the rest of the organism, which has its own nervous organization.

In the octopus's case there is a conductor, the central brain. But the players it conducts are jazz players, inclined to improvisation, who will accept only so much direction. Or perhaps they are players who receive only rough, general instructions from the conductor, who trusts them to play something that works.

5

MAKING COLORS

The Giant Cuttlefish

In the first chapter we met an animal hovering under a ledge in the ocean. As it hovered, it changed colors, second by second. An initial dark red unveiled patches of gray and silver veins. Blues and greens seeped back and forth on the arms. In this chapter we're back in the water with this animal and its ceaseless transformations.

A giant cuttlefish looks like an octopus attached to a hovercraft. It has a back shaped a bit like a turtle shell, a prominent head, and eight arms coming straight from the head. The arms are roughly like octopus arms—flexible and unjointed, with suckers. When you're facing a cuttlefish, these arms can look like they are placed in a roughly horizontal array, but they're arranged around the mouth, and like an octopus's arms, they can be thought of as eight huge and dexterous lips. Tucked away near the mouth are two longer "feeding tentacles," which can be whipped out to seize prey. The mouth itself contains a hard beak. A cuttlefish body has no spine or real bones, but there is a stiff "cuttlebone," which looks like the inside of a surfboard, inside the shield-like back. The shield is fringed by a skirt-like fin, a few inches wide, on each side. A cuttlefish moves slowly by undulating these fins. When it wants to move quickly it uses jet propulsion, with a "siphon" underneath the body that can be pointed in any direction. Most cuttlefish are small, measured in inches. But a giant cuttlefish can grow to three feet long.

This animal is three feet long with a skin that can appear just about any color at all and can change in seconds, sometimes much faster than a second. Thin silver lines wander over its head, as if the animal is visibly electrified. The electric lines make the cuttlefish look like a hovering spacecraft. But the disruption to one's impressions, to all attempts to make sense of the animal, is continual. As you watch, bright red trails lead from its eyes. A spaceship crying tears of blood?

Cephalopods in general (not all, but a great many) are skilled color changers. In this prodigious group, giant cuttlefish are perhaps the pinnacle, or at least the most colorful. Some degree of color change is not rare in nature; many animals can modulate their surface color to some extent. Chameleons are the familiar example. But cephalopods are faster and produce a wider range of colors. In the case of large cuttlefish, the entire body is a screen on which patterns are played. The patterns are not just a series of snapshots, but moving shapes, like stripes and clouds. These seem to be immensely *expressive* animals, animals with a lot to say. If so, what is being said, and to whom?

The giant cuttlefish is also remarkable in another way: how disarming it is to find friendliness in a large wild animal. I do not mean a mere toleration of a human's presence, but an active engagement, the animal's making contact with a foreign being. This is not routine in giant cuttlefish, but not rare either. Quite often you encounter a friendly curiosity. The animal comes forward, with its skin in a quiet "resting" pattern of colors and shapes. The cuttlefish will hover close by, apparently trying to work you out.

These are little-studied animals. They've not been kept in captivity very often. Alexandra Schnell, one of the few people to study them closely in the lab, says that they do show hints of the same complex responses to captivity seen in octopuses. They ambush visitors with well-aimed squirts of water from their jets. But giant cuttlefish seem even more enigmatic and otherworldly than their octopus relatives. They have big brains, both in absolute terms (sheer size) and as a proportion of body mass. As far as I know, they've not shown the most striking marks of intelligence seen in some octopuses—puzzle solving, the use of tools, the exploration of objects. But they have not been studied nearly as much, and their lives would seem to make such behaviors less useful than they are for octopuses. These are not clambering explorers, but swimmers.

While giant cuttlefish might not have the protean inventiveness of the octopus, they do have features that stay with you long after you've been around one in the sea: the friendly curiosity, at least in some cases, or a wary engagement as they hover toward you and away. And those unending, astounding color changes.

~ Making Colors

The skin of a cephalopod is a layered screen controlled directly by the brain. Neurons reach from the brain through the body into the skin, where they control muscles. The muscles, in turn, control millions of pixel-like sacs of color. A cuttlefish senses or decides something, and its color changes in an instant.

Here is how it works. The skin has an outer layer, a *dermis*, that acts as a covering. The next layer down contains the *chromatophores*, the most important of the color-control devices. A single chromatophore unit contains several different kinds of cells. One cell holds a sac of a colored chemical. Around it are muscle cells, one or two dozen of them, which pull the sac into different shapes. Those muscles are controlled by the brain. They stretch the sac to make its color visible, or relax it for the opposite effect.

Each chromatophore contains just one color. Different cephalopod species use different colors, and usually the animal has three kinds. In a giant cuttlefish, the chromatophores are red, yellow, and black/brown. Each is much less than a millimeter in diameter.

This device explains how cephalopods produce some of their colors, but not all. A giant cuttlefish can make red or yellow by activating chromatophores of one color alone, and it might make orange with a combination of the two. But this mechanism has no means to produce many other colors a cuttlefish might display. There's no way to produce blue, green, violet, or silver-white. Those colors are produced by mechanisms in the next layers of skin. Here we find several kinds of *reflecting cells*. These cells do not display fixed pigments, as chromatophores do, but reflect back incoming light. This reflecting need not be a simple mirroring. In *iridophores*, light is bounced and filtered through tiny stacks of plates. These plates separate and direct the light's different wavelengths, shining back colors that can be different from those that came in. The results include the greens and blues that chromatophores cannot produce. These cells are not attached directly to the brain, but it seems that some of them are controlled, more slowly, by other chemical signals. Just below the *iridophores*, the *leucophores* are another kind of reflecting cell; they do not manipulate the light but reflect it straight back. As a result, they often appear white, though they can reflect whatever color is around. As the chromatophores sit in a higher layer than the reflecting cells, all the reflecting cells have their effects modulated by what the chromatophores are doing. When chromatophores expand, this affects the light that makes it down to the reflecting cells, and hence what is shone back.

Imagine looking at a cuttlefish's skin from its side, in a cross-section. We would see a top layer, then a layer with millions of tiny colored sacs, each being pulled constantly into shapes that expose or hide the pigments inside. This will be happening at a great rate, through the activity of many muscles. Some light would pass through this layer and reach another layer, where it would be reflected and filtered between stacks of mirrors. Those cells might be changing their shape, more slowly, as chemicals reach them from elsewhere. Further down, a layer of simpler reflecting cells mirror back whatever light reaches them.

Here's a sketch of those layers:

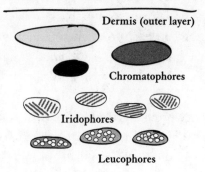

Suppose a giant cuttlefish has about ten million chromatophores. Then, very roughly, we can think of that layer of the animal's skin as a ten-megapixel screen. Roughly, I said, both because the pixels seem not to be controlled entirely independently of each other, but in local clumps, and also because each chromatophore has just one color. Some of the pixels are also on top of others, so the same patch of skin can produce many different colors. The layers below the chromatophores then add more complexity.

The cephalopod's colored layers are thin and fragile. Cuttlefish look very different when they've lost skin through age or damage. Then you see dull white patches. The magic skin is a thin sheet on top of a plain white body.

On the animals I watch, reds are in some sense the "base" colors, the most commonly seen. These reds range from maroon to fire-engine. A common decoration, on top of the red base, is a white-silver, as it appears underwater. The whites form veins and dots, making little jagged flashes or a line of pearls. Other colors appear in patches—yellows, oranges, olive-greens. They may hold static patterns but their colors are rarely fixed for long. Their "dynamic" patterns are like movies played on the screen of the cuttlefish's skin. An example is the *passing cloud* display. Alternating waves of darker and lighter patches move consistently along the body from front to back or back to front. Once I watched a large cuttlefish from above and saw the left side of its body displaying a passing cloud to another cuttlefish under a rock, while the right side was still and camouflaged, pointing out to sea.

Cuttlefish's color changes often occur in combination with changes to the shape of their body and skin. Sometimes they swim around with dozens of "papillae," or folds of skin, sticking straight out from their back. These look like tiny versions, an inch or so high, of the plates on the back of a stegosaurus. These papillae have nothing hard inside them, and can be produced in a second. The eyes are the sites of finely detailed modifications in skin shape. Many cuttlefish produce thin wisps and folds of skin above each eye. These look like carefully sculpted eyebrow extensions.

Octopus tetricus, or the "gloomy octopus," with arms roaming over its head. All the octopus photographs in the book are of this species, which is found in Australia and New Zealand.

This octopus has produced a very close color match to the seaweed behind it.

The next four images are video frames from a fight between two octopuses at the Octopolis site in Australia.

The vanquished octopus disentangles himself and jets away.

An octopus moving under jet propulsion, from right to left. This is the same animal who won the fight depicted on the previous pages.

An Australian giant cuttlefish, *Sepia apama*. This is Kandinsky, described in chapter 5.

This giant cuttlefish is showing early signs of age-related decline around his face and arms.

Rodin, a giant cuttlefish who spent a lot of time holding static poses with raised arms.

The eye of a giant cuttlefish has a pupil shaped like a *w*. Chromatophores—tiny sacs of pigment controlled by muscles in the skin—are visible around the eye. (This is the only photograph in the book taken with added light.)

These two photographs, taken four seconds apart, show a color change from dark yellow to red.

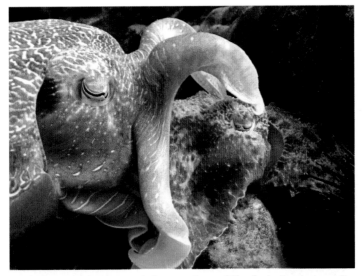

Two giant cuttlefish in a prelude to mating at Whyalla in South Australia, with the male on the left. There has been some scientific discussion about whether these animals are of the same species as those shown in my other cuttlefish photos, which were taken around Sydney. At least for now, only one species is officially recognized: *Sepia apama*.

This photograph, at Whyalla, shows the great range of colors giant cuttlefish produce using mechanisms layered in their skin.

A large and friendly giant cuttlefish swims alongside Karina Hall, who studies these animals and has taught me a lot about them.

A giant cuttlefish producing a complex range of reds, oranges, and silver-white marks. Both animals on this page are also shaping folds of skin above their eyes into temporary shapes.

At rest, the eight arms of a cuttlefish hang down in front and look fairly similar to one another. A cephalopod's arms have been assigned numbers, from 1 to 4, left and right. Starting at the top, there are arms left-1 and right-1. From the front these look like "inner" arms. Outside these are arms left-2 and right-2, then the third pair, and finally the fourth. In giant cuttlefish, the fourth arms are larger in males than in females. When showing aggression, males will often flatten their fourth arms into shapes like broad blades.

Another aggressive gesture is to hold the two "first" arms up like horns. Some cuttlefish make these horns elegantly wavy. Others shape their arms into fiddleheads, hooks, or clubs. In the most elaborate cases, cuttlefish will arrange layers of arms at three or four different levels. The first arms will be held high and straight; the second arms will be horn-like at a lower level, perhaps with curled ends, with the third pair below, and finally the fourth arms, flattened and made as massive as possible. There are some fish whom, despite their harmlessness, giant cuttlefish seem to positively despise, and their approach is generally greeted with arms raised into horns and hooks.

All these behaviors vary across individuals. I've sometimes been able to recognize an individual over many days, occasionally over a week. It's not easy to re-identify animals who can change their entire color and shape at will, but sometimes a distinctive scar makes this possible. Eventually I also learned to treat some white markings on the skirt-like fin, which seem to be permanent, as a kind of fingerprint. Different individuals respond differently to me though they are of the same sex and size, and in the same place at the same time of year. The most welcome style of interaction is the curious friendliness I mentioned above. Some individuals tend to come forward, in a resting pattern of colors, and watch closely. The friendliest of them have reached an arm forward to touch me. This is quite rare. The cuttlefish hovers, moving slightly with his fins or invisible jet. As we hover he maintains a specific distance, edging back when I edge forward, and sometimes forward when I edge back. But eventually, he lets the distance narrow until our bodies are only a few feet away. I move a hand out close to his arms, but do not touch them. The cuttlefish reaches the tip of an arm or two out to touch mine.

Nearly every time this has happened, it has happened only once. After a brief touch the cuttlefish shifts back to maintaining a few feet of distance. He was interested enough to touch, but having touched once he returns to where he was. One possible interpretation of this action is that the cuttlefish is seeing whether I might be good to eat. But a person is much larger than a cuttlefish, whose usual foods are crabs and fish caught whole. I don't think they are interested in me as lunch.

Some individuals, friendly or not, have distinctive styles of color change. I've occasionally encountered cuttlefish who seemed to produce colors the others had not quite thought of, or patterns with particular brilliance. The first of these I named Matisse. He was a friendly cuttlefish I visited for several days some years ago. All his colors had particular detail, but something else set him apart. He would be hovering without fuss in some mixture of reds and whites, and would suddenly explode into bright yellow. This flood covered his entire body, with no other marks visible, and would be switched on in less than a second. One moment he would be dark red, with veins and stripes, and in less than a second he looked like a cuttlefish-shaped sun. The flare would then fade, more slowly. Oranges would appear among the yellow, and darken. Patterning would return. In ten seconds or so he was dark red again.

The change to yellow was not accompanied by raised arms

or other displays; there was no other sign of fuss. I have seen "overall yellow" described as a sign of alarm in other cephalopods. I suppose it is possible that Matisse was alarmed, but why did everything else about him look so calm? Occasionally he produced yellow patterns in response to intrusive fish, but these were deeper yellows combined with arrangements of arms. The uniform blazes of canary yellow I saw appeared to be part of a different behavior. He just seemed to have a liking for these chromatic explosions.

In the years since I have seen quite a few other cuttlefish who produced these "yellow flares," though none who lit the water up quite like Matisse. Given what I said about the machinery of color change, it's easy to work out how this is done. A giant cuttlefish has some yellow chromatophores, so the flares were almost certainly produced by expanding, suddenly, those chromatophores and scaling back everything else.

Well after Matisse had come and gone, a cuttlefish arrived whose displays were beyond anything I have encountered. The only suitable name for him was Kandinsky.

Kandinsky had fixed habits and a definite home. Unlike Matisse, he did not have a single notable color. He produced the same kinds of patterns and colors that the others did, but in a more extravagant way. For about a week in 2009, when I was trying to get the perfect photograph of him, I would visit him in his home. I would arrive in the late afternoon each day and wait at the surface above his den, which was about twelve feet down. He would eventually emerge and come up to man the top of his rock, facing out to the ocean side. He held two arms aloft, with the others roaming about below them. I would swim down to meet him.

When I arrived he would have arms going everywhere, like a collection of ceremonial lances. Sometimes he would knot a

couple of arms together over his head. Raised arms are often
a sign of agitation and sometimes hostility, but I don't think this
was true of Kandinsky, as he tended to produce such elaborate
shapes continually, even when I was a fair distance away. On his
skin Kandinsky favored flaring mixtures of red and orange, including a kind of pale orange-green, and he often combined
these colors with the "passing cloud" display, in which waves of
dark shapes flow over the skin. Tear-like patterns moved down a
pair of his inner arms. After hovering in the water near his
favorite rock for a while, he would set off to tour the shallows.
He was not one of the friendly cuttlefish, but he would allow me
to follow closely as he roamed in loops through the reefs around
his den.

While some cuttlefish seem friendly and curious, a second
style of response to an exploring diver is intense hostility. Fortunately, this is rarer than friendliness. The most spectacular case I
remember was an encounter with a large male cuttlefish in a spot
where some very friendly ones have lived. Whenever I turn up at
that rocky ledge, I am reminded of those amiable meetings. This
time I found a perfect expression of animosity, choreographed in
color and shape.

I arrived and saw first a swirl of arms under the rock ledge.
The arms were yellow-orange-brown. The animal was facing out,
surrounded by waving seaweed, with arms going everywhere. I
thought initially this behavior might be camouflage—that his
waving was matched to the seaweed. I came closer and found
him producing more colors—silver-white welts. These were not
the relaxed pulsations of silver around the face and arms that are
common, but larger blotches that would flare on and off. His
lower arms were fanned out below, and the other arms were a
forest of horns. He was instantly watchful, and very soon he was
coming out at me, fast. I swam backward in a hurry. He kept

coming for some distance and then left the chase and returned to the den. I waited, and then approached cautiously again. Out he came, like a jet-propelled medieval siege device.

On these chases he produced the most murderous-looking displays I have ever seen: burning orange colors, arms like horns and sickles, and skin-folds resembling bent iron armor. Sometimes his inner arms were held high, contorted. At one point he held nearly all his arms aloft and twisted together, with just one set of arms below and his face between. I thought: he looks like the jaws of hell. It was as if he in his molluscan way had a real sense of what is frightening for a human, and was trying to produce a vision of damnation, something intended to strike at our hearts.

I persisted with him, and kept cautiously returning. He continued to chase me off, but I soon noticed that these charges never quite reached me, and this remained so when I began to retreat more slowly. I wondered how much bluff there was in his advances and how much genuine violent intent. Eventually I tried a new tack. If he is waving his arms murderously at me, why not wave back? The next time he came out I retreated much less, and lifted my arms in front of me, with scuba gear going everywhere. That got his attention. He still made *as if* to come forward, but did not actually move much, and the flailing arms began to calm down. He reduced his displays more and more, and soon his arms were at rest and the spiky skin-folds receded. I was finally able to come close to him. He stopped facing straight at me, and seemed to be looking at an angle, off over my shoulder, much more relaxed. Once I moved directly in front of him and he suddenly came at me again, his head lowered at first and then roiling with arms. I decided that we'd become as friendly as we were going to get.

There is another notable mode of human-cuttlefish interaction, though "interaction" is not quite the right word. Some cuttlefish behave with a level of indifference that is so intense it is hard to describe. In some ways this is the most intriguing behavior. These cuttlefish do not seem to register you as a living being at all. When they are still, they tend not to face a person directly (as others quite often do), but look off past one's shoulder. If you move a little, they adjust too. There is a maintenance of the non-contact.

This profound indifference is seen in some cuttlefish as they go on looping excursions around their reefs. On these trips a cuttlefish might poke around under rocks or just wander about. Much of the time they are probably looking for food or mates, but they often do not seem to be looking very hard. Touring cuttlefish can sometimes be friendly or at least curious, stopping to peer at you before swimming on. But some are able to ignore you no matter how close you swim—even if you are right alongside their eye. Once I was being ignored so perfectly that I planted myself directly in the animal's path, just to see what he would do. What followed felt like an existentialist game of "chicken." He came closer and closer, refusing to acknowledge my presence, until he was just a foot or so away. Then he looked up at me, with an expression that I cannot describe at all except to say that he seemed deeply unimpressed, edged past and swam on.

What role do we have then? What are we to them? Surely we are registered as large, mobile creatures. Surely, then, we might be potentially dangerous, or at least something of interest? Other cuttlefish do see us that way—as visitors to study, or to chase off with a wild display. But sometimes it appears that we do not come across as living beings at all. Being ignored so deeply makes you wonder if you are entirely real in their watery world, as if you are one of those ghosts who does not realize they are a ghost.

~ *Seeing Colors*

With our picture of cephalopod colors nearly complete, we now reach a fact that makes absolutely no sense. Cephalopods, in almost all cases, are said to be color-blind.

This impossible conclusion is based on both physiological and behavioral evidence. First, any system for detecting color differences requires something in the eye that can distinguish differences in the *brightness* of light from differences in its *color*. The usual way this is done is to have several different kinds of *photoreceptors*. Photoreceptor cells contain molecules that change their shape when they are hit by light. The change in shape triggers other chemical events in the cell; photoreceptors are the interface between the world of light and the signaling network of the brain. Any eye must contain something like this. For color vision, you need to have a range of photoreceptors that respond differently to the different wavelengths of the light that comes in. Most humans have three kinds of photoreceptors. Color vision—using this system—requires at least two. Most cephalopods have only one.

Behavioral tests have also been done on some species. Can a cephalopod learn to make a distinction between two stimuli that differ only in color, and in no other way? Apparently the ones that have been tested cannot.

This is baffling. These animals are *doing* so much with color. They are also superb at matching the color of their surroundings, for camouflage. How can you match colors you cannot see? Biologists sometimes offer explanations along these lines: First, cephalopods may be using subtle differences in brightness as indicators of the likely colors (hues) of objects around them, given the typical colors in their environment. Second, the reflecting cells, the mirrors in the skin, can help. You can produce a color you cannot see by reflecting it back from outside.

This makes sense of some of what cephalopods can do. Camouflage can be achieved with reflection—*if* the color you aim to match in your background is also coming toward you from other directions. Simple reflection cannot be the explanation if an animal is matching a color behind its back, while the light coming in from the front is different. In that case, the cephalopod would have to actively produce the right color—through some combination of chromatophores and reflecting cells—and it would have to know which color to produce. Cephalopods do seem to be able to do this; they often seem to match a color that is behind them when different colors are in front.

During the period I was writing this book, some pieces of this puzzle began to fall into place. The first pieces were put down in 2010, when Lydia Mäthger, Steven Roberts, and Roger Hanlon published a paper reporting that the photoreceptor molecules in the eyes of one kind of cuttlefish are probably also present in the cuttlefish's *skin*. That alone does not show much, for several reasons. First, it's possible that these molecules do something unrelated to seeing when they're found outside the eyes. Second, even if the light-sensitive molecules in the skin were indeed responding to light, this would not solve the color vision problem: there is still just one kind of photoreceptor molecule in the animal, even if it appears in strange places. You can't see color, it's thought, with only one photoreceptor.

For some years after the Mäthger-Roberts-Hanlon result was published, there was little follow-up. Through the Internet, I found just one person who seemed to be working on it: Desmond Ramirez, a graduate student in California. When I reached him, he confirmed that he was working on the problem, but played his cards close to his chest. Another couple of years passed. I'd just sent off a book review in which I wondered why the old lead was not being followed, and just days later, Ramirez published his paper. The paper, written with Todd Oakley, showed first that photoreceptor genes are active in the skin of a particular octopus species (*Octopus bimaculoides*). Crucially, it also showed that the skin of this octopus is sensitive to light and can change the shape of the chromatophores, even when the skin is detached from the body. Octopus skin itself can both *sense* light and also produce a *response* that affects the skin's color. Back in chapter 3 I discussed the way that an octopus's nervous system is spread throughout much of its body. The image I tried to develop in that chapter was one of a body that *is* its own controller, to some extent, rather than a body steered around by the brain. Now we learn that an octopus can see with its skin. The skin is not only affected by light—something true of quite a few animals— but it responds by changing its own delicate, pixel-like color-controlling machinery.

What could it be like to see with your skin? There could be no focusing of an image. Only general changes and washes of light could be detected. We don't yet know whether the skin's sensing is communicated to the brain, or whether the information remains local. Both possibilities stretch the imagination. If the skin's sensing is carried to the brain, then the animal's visual sensitivity would extend in all directions, beyond where the eyes can reach. If the skin's sensing does not reach the brain, then each arm might see for itself, and keep what it sees to itself.

The Ramirez and Oakley finding is an important development, but it does not yet solve the problem I emphasized above, the problem of color perception. The photoreceptor in the skin of Ramirez and Oakley's octopuses is sensitive to the same wavelengths as the photoreceptor in the eye. Even if the whole body can see, it seems that it must be seeing in monochrome. The problem of color match remains. I suspect, though, that the Ramirez work will lead to a resolution of this problem. A hint was offered in the older paper by Mäthger and her colleagues. They noted that even if the photoreceptors in the skin are chemically the same as the ones in the eye, their light sensing might be modulated by the chromatophores, or other cells, around them. This might permit one kind of photoreceptor to behave like two. Some butterflies use a similar trick.

This could work in several ways. One possibility is that a chromatophore might sit on top of a light-sensitive cell, acting like a filter. That photoreceptor would then respond to colored light differently from a photoreceptor paired with a differently colored chromatophore. Another possibility was suggested to me by Lou Jost, an ecologist, orchid expert, and artist. He suggested that the act of changing colors might do the trick. Suppose some light-sensitive cells sit below a layer of many chromatophores. As chromatophores of different colors expanded and contracted, the light passing through them would be affected in different ways. If the animal kept track of which chromatophores were expanded, as well as how much light was reaching its sensors, it could know something about the color of incoming light. The animal would be like a cameraman exchanging one filter for another, as it went through its color changes. A monochrome sensor can detect color if the organism has filters of different colors and knows which ones are operating at each moment.

All these possibilities depend on the location of the light-sensitive cells in relation to chromatophores, and on other unknowns. But in some ways it would be surprising if one of these mechanisms was *not* operating. As long as some light-sensitive structures lie below the colored chromatophores, when the animal performs its chromatophore changes there will inevitably be effects on the light-sensitive structures below, and these effects will correlate with the color of incoming light. The information is available. It would not seem to be a difficult evolutionary transition for the animal to put this information to use.

~ *Being Seen*

When it comes to camouflage, octopuses are unsurpassed. They can be completely invisible to an observer—an observer looking for octopuses—just a few feet away. They are helped by the fact that unlike cuttlefish, octopuses have almost no hard parts in their bodies, and can assume just about any shape at all. Giant cuttlefish cannot fool observers as completely as octopuses can, but some cuttlefish come close. The best act of cuttlefish camouflage I have seen was the work of a "reaper cuttlefish." This is a smaller species, growing to about six inches long. The grim name is quite misleading, as these are the sweetest-looking animals imaginable. They are usually a soft red with yellow eyeliner. I found this one in some seaweed. Once we saw each other he was very wary. He evaded me, swimming through weeds and around rocks, keeping obstacles between us. At one point he disappeared into a flat channel with a few rocks strewn around. In a moment I could not see him.

I knew that these cuttlefish can take on a mottled rock-like appearance, so I was positively expecting to find him somewhere trying to be a rock. There was a small rock in the middle of the channel. I looked and thought: well, that one is just a rock. I went to the other end of the channel where he should have come out, and there was no sign of him. I came back around to look at the channel again. And at the rock. Looking closer, it was the cuttlefish. Once it was apparent that I was fixated on him, he gave up the rock camouflage and wandered back into his dark pink. So there I was, looking for a small cuttlefish looking like a rock, in that exact spot, but he fooled me anyway.

Suddenly then, as I watched him change colors, a green moray eel raced out with jaws open and attacked him. There was a burst of ink from the cuttlefish—they have ink of the same sort as octopuses and squid. It looked like a cloud of black smoke, as if he had caught fire. I tried to see into the channel, which was now black, and caught only a brief glimpse of the cuttlefish being helplessly shaken and swept about by the eel. I felt terrible, as it seemed likely that I had distracted the cuttlefish and allowed the eel his chance.

Ink continued to billow out. Given the violence of the eel's attack, I had soon given up on the cuttlefish. But then he zoomed up out of the black cloud, wildly colored and strangely flattened, with his fins fanned out. He looked dazed, damaged, but still able to swim. He had just one large bite mark on the back of his body, and was still wearing his yellow eyeliner. At first he swam in a chaotic, punch-drunk meander. Then he straightened up and headed down toward another ledge.

I was amazed to see him. I think of a moray eel as a consummate predator, especially in that close-range context among rocks and weeds. They are all teeth, muscles, and snake-like strength. Once the eel was on him, it seemed there was no contest. The cuttlefish had no teeth or bones or armor. Rather than a flattened serpent, it looked like a toy hovercraft. Yet the cuttlefish escaped.

The original function of cephalopod color change—the reason it evolved—is thought to be camouflage. As cephalopods gave up their shells and began prowling waters full of sharp-toothed fish, camouflage was one way they avoided being eaten. Camouflage is the opposite of signaling; it's producing colors in order *not* to be seen or recognized. In some species, signaling then arose—the camouflage machinery was pressed into service as a way of communicating and broadcasting. Colors and patterns were now produced to be seen and noted by observers, such as rivals or potential mates.

Intermediate between the clear cases of camouflage and of signaling are *deimatic* displays. These are dramatic patterns often produced while fleeing a predator. It's hypothesized that they are an attempt to startle or confuse the foe—to suddenly look different, and weird, in a way that might lead the predator to pause or lose their bearings. Here, the display is supposed to be noticed, but it does not send information to a receiver. It is merely supposed to be confusing or disruptive.

During their mating period, giant cuttlefish males engage in ritualized displays that include an elaborate mix of skin displays and body contortions. This is seen most dramatically in one place on the southern coast of Australia, near an industrial town called Whyalla. Here thousands of giant cuttlefish gather off the coast each winter to mate and lay eggs in shallow water. No one knows why they choose this particular spot, but it's a great place to watch the most dramatic of all cephalopod signaling.

A large male will try to act as a "consort" to a female, monopolizing her and keeping other males away. When a rival male approaches, the consort and the intruder begin competitive displays. The two males will lie side by side quite close in the water. Each will stretch out as far as it can, often with a gentle curve in its body. They will blaze with color changes and patterns. Having stretched one way, a cuttlefish will often turn 180 degrees around and stretch out in the other direction. This turning, unfussed and deliberate, looks like a dance from the court of some civilized French king. The stretching, in contrast, looks like competitive yoga.

The mix of yoga and courtly dance suffices to determine which cuttlefish is larger, and the larger almost always prevails.

The smaller one will back off. The female will drift quietly in the water, perhaps staying close to her pulsating companion, perhaps wandering away. Cuttlefish sex, if it results, is a peaceful affair by the standards of the animal kingdom. They mate head to head. The male attempts to grasp the female front-on. If she accepts him, he will envelop the female's head with his arms. Having reached this position, there are a couple of minutes of stillness. Apparently he is blowing water at her with his funnel during this period. The male then uses his left fourth arm to take a sperm packet and place it in a special receptacle below the female's beak, and, with more rapid motions, he breaks the packet open. They separate.

Squid, also, engage in a considerable amount of signaling, much of it complicated and puzzling in its role. Some signals are clear, and common to several species. When a male approaches a female, she will sometimes display a definite white stripe that says "No thanks." In a moment I'll say more about some of these signaling systems, but first I want to outline something else I've come to believe about cuttlefish colors.

Let's accept that camouflage and signaling are the two *functions* of color change in cephalopods—it's due to them that color change evolved and is maintained. The fact that those are the functions of color change does not mean that every color you see is produced as a signal or as camouflage. I think that some cephalopods, especially cuttlefish, have an expressiveness that goes beyond anything with a biological function. Many patterns seem to be anything but camouflage, and are also produced when no obvious "receiver" of the signal is around. Some cuttlefish, and a few octopuses, go through an almost continual, kaleidoscopic process of color change that appears disconnected from anything going on outside them, and appears instead to be an inadvertent expression of the electrochemical tumult *inside* them. Once the color-making machinery on the skin is wired to the electrical network of the brain, all sorts of colors and patterns might be produced that are simply side effects of what is going on within.

That is how I interpret the colors of many giant cuttlefish; they are an inadvertent expression of the animal's inner processes. Such patterns include flares and surges of activity, and also subtler changes. If you look closely at the "face" of a giant cuttlefish—the area between its eyes and down the first part of its arms—you will often see an ongoing murmur of very small color changes. Perhaps the machinery of color change is in an "idling" state there. I spent several days visiting a cuttlefish I called Brancusi. He rarely produced bright colors. Instead he would sometimes fix a few of his arms into an unusual pattern and then hold the shape completely still, like a sculpture, for as long as I was able to stay with him. He would hold up a pair of inside arms like horns, but angle the top of them down toward the sea floor. Brancusi favored shape rather than color, but if I looked closely, I would see a constant restlessness in all the colors on his face. In other animals I've often seen steady pulsing changes just beneath the eyes, like animated eye shadow.

I accept that cuttlefish can control their skin closely when they want to. They can snap into camouflage, or an aggressive display, very quickly. Any color changes that do not contribute to signaling and camouflage are side effects, from an evolutionary point of view. If they did much harm they would probably be suppressed. But perhaps they do not do much harm. More precisely, perhaps they would do harm—attracting unwanted attention—in smaller cephalopods, but do not do much harm in a giant cuttlefish, an animal big enough for many predators to pass by.

Another possibility is connected to the speculative ideas

about color sensing I described earlier. Suppose that by changing its colors a cephalopod affects the light that reaches sensors within its skin. Then some of these ongoing low-level color changes might be a way of surveying the chromatic environment.

I realize also that a lot of the color changes that puzzle me might be triggered by my own presence. I often try to keep some distance away and off to the side when watching these displays. I have also set up video cameras at octopus dens and then left for a few hours, to see what they do when no one is around. The animals often go through unexplained sequences of colors even when, as far as I can tell, no other octopuses are nearby. Perhaps the camera is their intended audience in these cases. That's possible. But another possibility is one that takes things more at face value. I think these animals have a sophisticated system designed for camouflage and signaling, but one that is connected to the brain in a way that leads to all sorts of strange expressive quirks—to a kind of ongoing chromatic chatter.

~ Baboon and Squid

Signals are sent and received; they are made to be seen or heard. To look more closely at sender-receiver relationships in animals, we'll emerge from the water and switch to a very different case. Wild baboons in the Okavango Delta of Botswana, Africa, have been studied for years by Dorothy Cheney and Robert Seyfarth, two of the most influential researchers into animal behavior.

The life of a baboon is a fraught one. There are constant risks from the great African predators, and also an intense, shifting social scene to contend with. Baboons live in troops. The one Cheney and Seyfarth studied contained about eighty individuals with a complicated dominance hierarchy. Female baboons remain in the troop they were born into and form a hierarchy of families (matrilines), with further dominance relations within each matriline. Most males leave the group of their birth and migrate into another as young adults. They live shorter and tougher lives, with more violence and many exhausting chases and displays. They are frequently driven off or drive others off. Even when the composition of a group is stable, both sexes face challenges and shifts, form alliances and friendships, and do a lot of grooming.

All this is meticulously documented by Cheney and Seyfarth in the book *Baboon Metaphysics*. Given the complex social life, it's not surprising that there is communication. But baboons can make only fairly simple sounds—calls of three or four kinds, especially threats, friendship grunts, and submissive screams. Communication itself is simple, but as Cheney and Seyfarth show, it gives rise to some sophisticated behaviors. Each individual calls in a distinctive way, and a baboon can recognize who has just called—they know *who* has threatened and who has backed off. Cheney, Seyfarth, and others worked out, by means of ingenious playback experiments, that a baboon hearing a series of calls is able to process it in very rich ways.

Suppose a baboon hears this sequence coming from a location it can't see: A threatens and B backs down. What does this mean? It depends who A and B are. If A is higher in the hierarchy than B, it's not surprising or notable. But if A is below B, then a sequence in which A threatens and B backs down is surprising and important. It indicates a change in the hierarchy, something that will matter to a great many members of the troop. In the playback experiments, a baboon would behave differently, being much more attentive, when a series of calls indicated an important event of this kind. As Cheney and Seyfarth say, it seems that the baboons construct a "narrative" from the series of sounds they hear. This is a tool they use for the purposes of social navigation.

Compare the baboons with the cephalopods. In baboons, the production side of their vocal communication system is very simple. There are only three or four calls. An individual's choices are limited, and a call will reliably follow interactions of a particular kind. The *interpretation* side, though, is complex, because calls are produced in ways that allow a narrative to be put together. The baboons have simple production, complex interpretation.

The cephalopods are the opposite. The production side is vastly, almost indefinitely complex, with millions of pixels on the skin and a huge number of patterns that might be produced at each moment. As a communication channel, the bandwidth of this system is extraordinary. You could say *anything* with it—if you had a way to encode the messages, and if anyone was listening. In cephalopods, though, social life is much less complicated than it is in baboons, as far as anyone can tell. (I will discuss some surprises below and in the last chapter, but they won't change this comparison—no one thinks any cephalopod has a social life comparable to that of a baboon.) Here we have a very powerful signal production system, but most of what is said is going unnoticed. Perhaps that's not the right way to put it: perhaps because no one is interpreting most of it, little is really being *said*. But it's also true that with all the chatter, all the mumbling of the skin, a lot of what is going on inside is made *available* on the outside.

Signal production in one cephalopod species, the Caribbean reef squid, was documented extensively back in the 1970s and 1980s by two researchers working in Panama, Martin Moynihan and Arcadio Rodaniche. They followed their animals in the field for years, recording their behaviors in detail. They found a lot of complexity in the patterns being produced, so much so that they suggested the squid have a visual *language*, with a grammar—with nouns and adjectives and so on. This was quite a radical claim to make. They published their ideas in a monograph that appeared in a very respectable journal, but it was an unusual publication, with personal reflections and ongoing attempts to get inside the world of these skittish animals, whom they patiently followed around all day with snorkels. The monograph was also beautifully illustrated by Rodaniche, who later retired from science and became an artist.

The argument for a visual language was made by showing the complexity of squid displays. These displays combined colors and body postures—some of them are tiny analogues of giant cuttlefish displays described above. Moynihan and Rodaniche charted the sequences they saw—*Golden Eyebrows, Dark Arms, Downward Pointing, Flecked Yellow, Upward Curl* . . . I once chased one of these squid over a reef in Belize, and was struck, as they were, by the complexity of what it was doing. But there is a mismatch in Moynihan and Rodaniche's own discussion, one that they were aware of but perhaps did not fully confront. Communication is a matter of sending and receiving, speaking and hearing, producing and interpreting—two complementary roles. Moynihan and Rodaniche were able to document a lot of very complicated production of signs, but they said much less about the signs' effects—how the patterns were being interpreted. They were able to work out a few fairly clear sign-and-response combinations in mating situations, but many of the displays they observed were produced outside that context.

In total they counted about thirty ritualized displays, and many patterns in the sequences and combinations of displays that were produced. They said that these patterns must have *some* meaning, but in most cases they could not work out what it was: "We cannot, ourselves, in the present state of our knowledge, always and in every case tell the difference in message or meaning between every observed arrangement of particular patterns. We feel, nevertheless, that we must assume that there is a real functional difference of some sort between any two sequences or combinations that can be distinguished from one another." By their own lights, there was not a lot of complexity in the behavioral interactions between one squid and another. Why, then, were such complex displays being produced?

There's a genuine puzzle here. Even if Moynihan and Rodaniche were overcounting the signals and making too much of the analogy with language, there's still a question about why the squid seem to be saying so much. It's possible that the sequences of colors, poses, and displays played various subtle social roles. Later researchers have been a bit skeptical about this part of Moynihan and Rodaniche's work. But perhaps there's more going on than we can tell.

These squid are among the most social of cephalopods. The contrast between the baboons and the cephalopods is, I hope, vivid. In the cephalopods we find, as a result of their heritage of camouflage, an immensely rich expressive capacity—a video screen is tied directly to their brain. Cuttlefish and other cephalopods are brimming with output. Publish or perish. To *some* extent, this output is designed by evolution to be seen; sometimes it is camouflage, but sometimes it is meant be noticed, by rivals and the opposite sex. The screen also seems to run through much chatter and murmuring, happenstance expression. And even if cephalopods have hidden powers of color perception, a lot of their wild chromatic output is surely lost on watchers. The baboons, on the other hand, can say hardly anything. Their channel of communication is very limited. But they hear much more.

These are both partial cases, *unfinished*, in a sense, though one should not think of evolution as goal-directed. Evolution is not heading anywhere, not toward us or anyone else. But I can't resist seeing, in both animals, an unfinished quality. They are both animals with a one-sidedness in their version of the fundamental signaling duality, the interlocking roles of sender and receiver, producer and interpreter. On the baboon side, there's a soap opera life, frantic and stressful social complexity, and little means to express it. On the cephalopod side, there's a simpler social life, hence less to say, but such extraordinary things expressed nonetheless.

~ Symphony

Late one summer afternoon I swam down on scuba to a favored place, a den where I have seen many giant cuttlefish. A cuttlefish was there. He was medium-size, probably male, and even from some distance I could see that he had intense colors. He did not mind my arrival, but was not curious or watchful. He was very still.

I set myself up next to him, just outside his den. As he faced out past me toward the sea, I watched as his colors changed. The series was mesmerizing. I noted right away a *rust* color, different from the reds and oranges one usually sees. You would think that every shade of red and orange had been covered a hundred times by animals I'd seen, but this one seemed unusual, a rust-with-brick. There were also gray-greens, other reds, and faint pale colors I could not quite catch.

As I watched, I realized that these colors were changing in a concerted way, and changing in more ways than I could track. It reminded me of music, of chords changing amid and over each other. He would shift several colors in sequence or together—I was not noticing which—and end up with a new pattern, a new combination, which might stay still for a time or immediately start shifting into another. There were dark-yellow–pale-brown combinations, red combinations that were more familiar, and others. What was he doing? It was slowly getting darker in the water, and under his ledge it was already quite dark. He was not displaying much with his body. I remained off to the side, as still as I could be and breathing as little as possible. The eye facing me seemed nearly closed, but I've learned that cuttlefish are able to see more than one might expect with their eyes mostly shut.

He looked out into the darkening sea, where yellow-green seaweed was waving. Given this motion, I wondered whether this might be a case of the "passive" production of colors, reflecting the mix coming in. But the movement through colors seemed more organized than that, and many of the colors had no analogues outside. He kept moving through his chords.

I crouched low among the weeds. It occurred to me that he was paying *so* little attention to me that all of this might have been going on while he was asleep, or half-asleep in a state of deep rest. Perhaps the part of his brain that controls the skin was turning over a sequence of colors of its own accord. I wondered if this was a cuttlefish dream—I was reminded of dogs dreaming, their paws moving while they make tiny yip-like sounds. He made almost no movement, except small adjustments of siphon and fin that kept him hovering in the same place. He seemed to be maintaining as little physical activity as possible, except for the ceaseless turnover of colors and patterns on his skin.

Then things started to change. He seemed to stiffen or pull together, and began going through a long series of displays. It was the strangest series I have seen, especially as it seemed to have no target or object. During almost all of the sequence he faced well away from me out to sea. He pulled in his arms and exposed his beak. He tucked his arms below him in a missile-like pose, then produced a yellow flare. I kept glancing out to see if he was looking at someone—another cuttlefish or some other intruder. There was never anyone there. At one point he went into the sideways stretch that males do when they are competing with each other. Then he pulled himself into the most extraordinarily contorted shape, his skin suddenly white, with arms pulled back both above and below his head. This sequence then quieted down. I backed off and moved higher in the water, remaining beside the den and not in front of it, and watching him calm down. And then, instantly, he seized into a wild aggressive pose, with arms straight out, sharp like thin swords, and his whole body a bright yellow-orange. It was as if the orchestra suddenly hit a wild clashing chord. The arms ended in needles, his body became covered with jagged papillae like armor. He then began roaming a little, sometimes facing me and sometimes facing out to sea. I wondered again if this was all directed at me, but if it was a display, it seemed to be aimed in directions all around. And I had been back from the den when this sequence started, when he exploded into yellow-orange and the needle-arm pose.

Still facing away, he began to ease back from this *fortissimo*. Though he moved through a few more permutations and poses, they were subsiding. And then he was still—his arms hanging down, his skin a quietly shifting mixture of the reds, rusts, and greens that he had been producing when I arrived. Turning, he looked at me.

I was now cold and the water was getting steadily darker. I had been there beside him for perhaps forty minutes. Now he was calm, and with the symphony or dream over, I swam in.

6

OUR MINDS AND OTHERS

From Hume to Vygotsky

In one of the most famous passages in all of philosophy, David Hume in 1739 looked inside his mind in an attempt to find his *self*. He tried to find some enduring presence, a permanent and stable being which persists through the jumbled flow of experience. He claimed he could find no such thing. All he could find was a rapid succession of images, momentary passions, and so on. "I always stumble on some particular perception or other, of heat or cold, light or shade, love or hatred, pain or pleasure. I never can catch *myself* at any time without a perception, and never can observe any thing but the perception." These sensations or perceptions, he said, comprise him—nothing more. A person is just a bundle or collection of images and feelings, "which succeed each other with an inconceivable rapidity, and are in a perpetual flux and movement."

Hume's look inside provides a good starting point for this chapter because everyone can do what he did. When we do, despite Hume's confident inventory, we surely find two things that he did not mention. First, Hume described what he found inside as a "succession" of sensations. But it seems more accurate to say that we find a *combination* of sensations present at each time. Our experience usually forms an integrated "scene"—a mix of visual information, sounds, a sense of where our body is, and so on. It's not one impression after another, but several at each time, tied together. As time moves forward, one such combination passes into another.

The second thing Hume missed is more conspicuous. When we look inside, most people find a flow of *inner speech*, a monologue that accompanies much of our conscious life. Sentences and phrases, exclamations, rambling commentaries, speeches we would like to give, or wish we'd given. Maybe Hume did not find this in his own case? Some people have a more prominent inner monologue than others. Perhaps Hume was one of those for whom inner speech is weak? It's possible, but I think it's more likely that Hume did encounter inner speech, but regarded it as one part of the wash of sensations, not as anything special. There are colors and shapes and emotions in there, and echoes of speech as well.

Hume's inattention to inner speech may also have been guided by his overall agenda in philosophy, by the *shape* of the theory he wanted to defend. Hume was inspired by Isaac Newton's theories in physics, unleashed about fifty years before. Newton saw the world as made up of tiny objects ruled by laws of motion and a principle of attraction between them, also known as gravity. Hume aimed at an explanation of the same kind of the contents of the mind, and thought that he had discovered a "power of attraction" between sensory impressions and ideas, a complement to Newton's attraction between physical objects. Hume wanted a science of the mind that took the form of a quasi-physics, in which ideas would behave like mental atoms. The peculiar properties of inner speech are of little relevance to this project, and the contents of Hume's personal mental stock-taking matched his philosophical goals rather well. Nearly two centuries after Hume, the American philosopher John Dewey, who saw the world very differently, remarked, "It is altogether likely that the 'ideas' which Hume found in constant flux whenever he looked within himself were a succession of words silently uttered."

Around the same time as Dewey's comment, a young psychologist was developing a new theory of thought and child development in the tumultuous early years of the Soviet Union. Lev Vygotsky grew up in what is now Belarus, the son of a banker. He had just finished his student years when the 1917 Russian Revolution broke out. He worked with the Bolsheviks in local government for a time, supported Marxist ideas, and developed his psychological theories in a Marxist context. Vygotsky thought that as a child developed, progressing from simple responses to complex thought, a transition takes place through the internalization of the medium of speech.

Ordinary speech, saying things and hearing them, plays an organizational role in our lives—it helps us put ideas together, draw attention to things, get actions to occur in the right order. Vygotsky thought that as children acquire their spoken language, they also acquire inner speech; a child's language "branches" into inner and outer forms. Inner speech for Vygotsky is not merely an unspoken version of ordinary speech, but something with its own patterns and rhythms. This inner tool makes possible organized thought.

Both physically and intellectually embedded in Soviet Russia, Vygotsky was not influential in the West. Around 1930, he suffered a personal and intellectual crisis and began revising his theories. He also found himself dealing with dangerous accusations of "bourgeois" elements in his work. Vygotsky died, at just thirty-seven, in 1934.

In 1962 an English translation of his work *Thought and Language* appeared. Vygotsky is still regarded as a bit of a fringe figure in psychology. A few prominent people working today, such as Michael Tomasello, acknowledge his influence (the first time I remember seeing Vygotsky's name was in an acknowledgment in a famous book by Tomasello), but many do not. With or without credit, the picture he sketched is becoming increasingly important as we try to understand the relations between human minds and others.

~ Word Made Flesh

What is the psychological role of language, our ability to speak and hear? In particular, what is the role of all that chattering, rambling internal speech? These questions prompt sharp divides. For some, inner speech is idle commentary, froth on the surface of the mind, and not very important. For others, like Vygotsky, it is a crucially important tool. In a brief but famous comment in *The Descent of Man*, of 1871, Charles Darwin claimed that speech, either inner or outer, is needed for complex thought.

> The mental powers in some early progenitor of man must have been more highly developed than in any existing ape, before even the most imperfect form of speech could have come into use; but we may confidently believe that the continued use and advancement of this power [of speech] would have reacted on the mind by enabling and encouraging it to carry on long trains of thought. A long and complex train of thought can no more be carried on without the aid of words, whether spoken or

silent, than a long calculation without the use of figures
or algebra.

Initially, this view might seem inevitable—that complex
thought, with its movement from premise to conclusion, step to
step, must require language or something close to it. It seems that
organized internal processing could not take place without it.

But once we make that last statement, we're saying something that's not true. It has become clear now that very complex
things go on inside other animals without the aid of speech.
Remember the baboons from the previous chapter. They live in
social groups with complex alliances and hierarchies. They have
simple vocalizations, three or four calls, but their internal processing of what they hear is much more complicated. They can
recognize each individual's calls and interpret a series of calls made
by different baboons, constructing an understanding of events
around them that is far more complex than anything a baboon
could *say*. When they build these narratives, they have some means
for putting ideas together which goes far beyond what they can
express using their communication system.

The baboon case is especially compelling, but there are
others. Recent years have seen a steady and surprising upgrading
in our view of what some kinds of birds can do, especially crows,
parrots, and food-storing birds like jays. Nicola Clayton and
others at the University of Cambridge, through a long series of
studies, have shown that birds can store food of different kinds
in hundreds of distinct places to retrieve later, and can remember not only where they have put food but *what* was put in each
place, so the more perishable items can be retrieved before the
longer-lasting ones.

Vygotsky himself, back in the early twentieth century, came
to recognize some of this. He knew of the first glimmers of

work showing complexity in animal thought, work that may have been quite disruptive to his theories. Vygotsky initially thought that the internalization of language must be essential to any sort of complex internal processing, but then he became aware of Wolfgang Köhler's work on chimps. Köhler was a German psychologist who spent four years working in a field station on Tenerife in the Canary Islands around the time of World War I. On his island he studied nine chimpanzees, and looked especially at how they gained access to food in novel situations. The chimps sometimes seemed to show "insight," Köhler said; they could work through novel problems spontaneously. Most famously, they stacked boxes on top of each other and climbed up on them to reach food hanging out of reach. Köhler weakened the idea that there is a necessary link between language and complex thought.

Some evidence pushes this way even in the human case. The Canadian psychologist Merlin Donald's book *Origins of the Modern Mind*, published in 1991, made use of two "natural experiments." First, he looked at evidence about the lives of deaf people in preliterate cultures that also lacked sign language. He argued that these people lived more normal lives than we'd expect if language was essential to complex thought. Second, he used the remarkable case of a French Canadian monk known as "Brother John," who was described in a 1980 paper by André Roch Lecours and Yves Joanette. Brother John lived normally most of the time but was subject to occasional attacks of severe aphasia. During these episodes he lost all use of language, both speech and comprehension, both public and internal. He remained conscious during these episodes, which sometimes occurred in a public setting, in which case he had to deal with them as inventively as possible. The paper describes one episode in which he arrived by train in a town, had one of his attacks, and had to find a hotel and order something to eat. He did this by means of gestures (including pointing at what he assumed was the right part of an illegible menu), and he did it without any internal linguistic stream to organize his thoughts and actions. If the view that language is essential for complex thought is true, Brother John should have been much less able to function than this. John would later describe these episodes as very difficult and confusing, but he did manage, and he was mentally *present* during them.

Extreme views on both sides of the question are fading: language is an important tool for thought, and inner speech is not mere mental-acoustic froth. But it is not essential to the organization of ideas, and language is not *the* medium of complex thought. I said in my opening paragraphs that Hume's inventory of inner life was surprising in its neglect of inner speech, but you might have had an exactly analogous response to the comment I quoted from John Dewey. Dewey reckoned that the "ideas" in Hume *were* just a series of words silently uttered. Even if words were indeed present, was Hume wrong to say he also found "heat or cold, light or shade, love or hatred"? Surely Dewey encountered such things in himself, too. Both philosophers' catalogues seem incomplete.

The role language plays in our minds might not be too different from the role Darwin sketched, though Darwin put it in too strong a form. Language provides a medium for the arrangement and manipulation of ideas. Here is an example from recent research on young children done in the laboratory of the Harvard psychologist Susan Carey. She looked at when children become able to use a logical principle called the *disjunctive syllogism*. Suppose you know that *either A or B* is true. Then you learn *not A*, so you should conclude *B*. Can children follow this rule before they have the word "or" in their vocabulary? For a while it was thought that they could, but now it seems they need to have learned the word before they can do this sort of mental processing. (If the sticker is under this cup or that cup, and you learn it's not under this cup, then . . .) It's always hard to work out the relation between cause and effect in studies of this kind, but this looks like a very Vygotskian result.

What are the inner mechanisms by which all this works? How is the word made flesh? There is a great deal of uncertainty here. But one plausible model, drawing on the work of several people, goes like this.

Ordinary speech functions both as input and as output. Hearing provides the mind with input; our speech is an output. We both speak and hear, and we can hear *what* we say. Even talking to yourself out loud can be a useful way of approaching a problem. I'll now tie these familiar facts to a concept that has become increasingly important in the brain sciences: the concept of an *efference copy*. (The word *efference* here means the same thing as output, or action.) The best way to introduce the idea is through the example of vision.

When you move your head or shift your gaze, the image on your retina continually changes, but this is not perceived as a change in the objects around you. You continually compensate for your own eye movements, so when something *does* move in the environment, you register it. This requires that you keep track of your own decisions to act. With an efference copy mechanism, as you decide to act, sending a "command" of some sort to your muscles, you also send a faint image of the same command (a "copy" of it, in a rough sense of that term) to the part of the brain that deals with visual input. This enables that part to take into account what your own motions are doing.

Without using the term, I introduced the idea of efference copies back in chapter 4, when I looked at how evolution created new kinds of loops between action and the senses. Mobile animals of many kinds have to deal with the fact that what they *do* affects what they *sense*; this creates the problem of distinguishing when a change in what is perceived is due to something important happening outside, and when it is due to the animal's own actions.

As well as helping solve problems of perception, these mechanisms also have a role in the performance of complicated actions themselves. When you decide to act, efference copies can be used to tell the brain, "Here is how things should look, given what I just did." If things don't look as expected, that might be because something in the environment has changed, but it might also be because the action you tried to perform did not come out as planned. You often need to work out whether your *trying* to do X resulted in your actually *doing* X. You know how things should feel if you push against a table, for example. If things don't feel the way you expect, that might mean the table was on rollers, but it might mean you did not succeed in pushing on the table at all.

Now let's apply all this to the case of speech. Everyone wants their words to come out as planned, and speech is a very complex action. In speech, the creation of an efference copy enables you to compare your spoken words to an inner image of them; this can be used to work out whether the sounds "came out right." As we say things out loud, we also register, internally, the sounds of what we *meant* to say, and we can then tell if the words came out incorrectly. Ordinary speech involves, in the background, a kind of internal quasi-saying and quasi-hearing.

So far, this hidden side to ordinary speech is helping with the control of complex actions. But these auditory images of speech, these internal quasi-said sentences, seem to have taken on other roles as well. Once we are generating these nearly-spoken sentences to check what we actually say, it's not too big a step to put together sentences that we *don't* intend to say, sentences and fragments of language that have a purely internal role. The forming of sentences in our auditory imagination creates a new medium, a new field of action. We can formulate sentences and experience their results. When we hear—internally—how some words hang together, we can learn something about how the corresponding *ideas* hang together. We can put things in order, bring possibilities together, can list and instruct and exhort.

Earlier, I mentioned John Dewey, who commented on Hume's omission of inner speech when he described what he found within. For Dewey, inner speech was important but its role was largely recreational, a vehicle for storytelling. It's odd that he did not discuss other uses. This might be because Dewey was so intensely social a philosopher; he thought that most of the important things we do take place out in the open. For Vygotsky, inner speech has a role in what is now called *executive control.* Inner speech gives us a way of performing actions in the right order (*turn off the power first, then unplug the machine*), and exerting top-down control over habits and whims (*don't eat another slice*). Inner speech can also be a medium of experiment, for putting ideas together to see what comes of the combination (*how would things look if I could travel as fast as light?*). Within a terminology used by Daniel Kahneman and other psychologists, it's a means for *System 2* thinking. This is a slow, deliberate style of thinking we engage in when we encounter novel situations, as opposed to the rapid *System 1* thinking that makes use of habits and intuitions. System 2 thinking tries to follow proper rules of reasoning, and tries to look at things from more than one side. It is ponderous but powerful. It is how we avoid temptation (if we do) and how we assess whether some novel action will actually get the job done.

Inner speech seems to be an important part of System 2 thinking. It is a way of walking through the consequences of actions, a way to bring reasons to bear against temptation. Gesturing to the careening inner monologues of the novels of James Joyce, Daniel Dennett has called the outcome of this wiring-in of speech a *Joycean machine* in our heads. But how could something as mundane as an efference copy system give rise to something so powerful? The mere existence of bits of language floating about inside us ought not to have so many consequences.

Part of the explanation may lie in the way that sentences of inner speech can be *attended* to. They are made available to much of the brain in something like the same way that ordinary speech is. Indeed, the similarities are so strong that it's easy for people to mistake sounds that exist only in their auditory imagination for sounds they are actually hearing. In an experiment done in 2001, people were told to listen to featureless random noise through a set of headphones, and were told that the song "White Christmas" might occasionally be played very quietly through the noise. They had to press a button if they were sure they'd heard the song. About a third of the subjects pressed the button at least once, but in fact the song was never played. The usual interpretation of the experiment is that the subjects in the experiment imagined the tune they were supposed to listen for, and sometimes mistook their own auditory image for a genuine playing of the song. The sounds we cook up in our heads, including the sounds of words, are *broadcast* in our minds in something like the way that many ordinary perceptual experiences are broadcast. Once a sentence of inner speech is composed, it is exposed to the same sort of processing that would apply to a sentence we hear. A novel combination of ideas, or an exhortation to act, is thus made available for consideration; it can have the same sort of effect that an ordinary spoken sentence can have. These phenomena, including the "White

Christmas" experiment, have informed some attempts to explain a common symptom of schizophrenia in which people "hear voices" in a way that disrupts action and a sense of self.

Inner speech is apparently one of a family of tools that enable complex thought in us. Another is spatial imagery, inner pictures and shapes. In landmark work from the 1970s, the British psychologists Alan Baddeley and Graham Hitch gave a model of *working memory*, a short-term store we all have for a few items of information that are being retained or worked on, usually consciously, from moment to moment. Baddeley and Hitch thought that working memory has three components: a *phonological loop*, which can play imagined sounds such as inner speech, a *visuospatial sketchpad*, which we use to manipulate pictures and shapes, and an executive control device that orchestrates the activities of the other two subsystems. Inner sketches and shapes are in some respects very different from inner speech, but they are also tools for complex thought, and may have similar origins in efference copy mechanisms—in this case, from the ways we control hand motions and gestures.

There is a lot missing from our knowledge in this area, and some major features of the picture I sketched are conjectural. The origin of inner speech and its relatives in efference copy mechanisms has not been demonstrated, but merely hypothesized. It's possible that inner speech and imagery instead have different origins. They may arise purely from the imagination itself, and only coincidentally resemble products of the ancient mechanisms that make possible complex actions.

~ *Conscious Experience*

Inner speech, and the sketches and shapes that inner language tangles with, have huge effects on subjective experience. Any

ordinary human has at his or her disposal a field for the performance of countless invisible actions. The echoes and commentary, the chatter and cajoling, are as vivid as anything in our inner
lives. You can be sitting motionless, watching an unchanging scene,
and your mind can be *alive* with this stuff, teeming with it in a
great jumble. Inner speech is so subjectively prominent, for many
people, that it can be overwhelming; the endless chatter is something people use meditation to get *away* from.

What do these features of human thought tell us about the
origins of subjective experience? In chapter 4, I sketched a framework in which the explanation comes in two parts. First, there are
basic forms of subjective experience that arise from widespread
features of animal life. I take pain to be an example. The second part of the story is about the evolution of more sophisticated kinds of subjective experience—*conscious* experience, in
a substantial sense of that term.

I think that inner speech and its relatives, the tools I've discussed in this chapter, can fill out this picture. In chapter 4
I introduced the workspace theory of consciousness, first proposed by the neurobiologist Bernard Baars. Baars tried to explain conscious thought in terms of an inner "global workspace,"
where lots of information can be brought together. As Baars
saw it, most of what happens in our brains goes on unconsciously, but a small fraction of it can be made conscious by being brought into the workspace.

When this idea was first proposed in the late 1980s it seemed
too close to old views that tried to explain consciousness by
finding a special *place* in the brain, a place where thoughts somehow acquire a subjective glow. Baars encouraged this spatial metaphor; the workspace is like center stage. I've seen people who
defend the workspace idea get into difficulty as a consequence
of this—"What makes the workspace special? Is there a little

man living there?" The workspace idea did look awkward when
it appeared, but Baars was onto something, and scientific work
guided by this idea soon bore that out.

Baars took as one of his starting points the idea that human
subjective experience is *integrated*. Information from several different senses, and from our memory, is brought together to give
us a sense of an overall "scene" that we inhabit and act in. A
second-generation version of the workspace theory was defended
by the French neurobiologists Stanislas Dehaene and Lionel
Naccache in 2001. Dehaene and Naccache argued that conscious
thought in humans has a special relation to novel situations and
actions that take us outside our routines. We start to deal with a
task consciously when our habits break down, or can't be applied,
and we have to do something new. Often, working out this new
action requires putting several different kinds of information together and seeing what comes of them. For Dehaene and Naccache, the function of conscious thought is to make it possible
for us to do novel, deliberate actions that require us to take the
"big picture" into account.

This approach is usually called the "workspace theory," but
there have always been two ways of talking about it, two metaphors that people call on to describe it. Baars, Dehaene, and
Naccache also talk about a kind of *broadcast* when describing
how consciousness works: broadcasting information throughout
the brain is what makes that information conscious. Sometimes
they talk as if both things are needed, a workspace and broadcasting (Baars talks like that); other times it seems that these are
two metaphors being used to help us understand a single thing.

I think the metaphors are very different, though, and "broadcast" is not even clearly a metaphor in this context. The idea of
integration by means of broadcast should be seen as a replacement
for the idea of an inner workspace, not as another way of ex-

pressing the same idea. "Where is the inner space? Who looks at it?"—these questions are not troubling when we use a model of broadcasting. From there, the next step is to see that inner speech and its relatives provide a *means* for broadcasting, a way it can be done. Inner speech provides one way we are able to route things through our minds in such a way that information can be assessed and used. Inner speech does not live in a little box in your brain; inner speech is *a way your brain creates a loop*, intertwining the construction of thoughts and the reception of them. And when that's done, the format provided by language allows you to bring ideas together in an organized structure.

I don't present this as a complete theory of inner broadcasting and its relation to conscious thought. Dehaene and other neuroscientists trace out mechanisms for the broadcast and integration of information that probably don't have anything to do with inner speech. I do think it's part of the story, though, and one of several ways in which efference copies and inner speech contribute to an explanation of the special features of human experience.

Here is another. A phenomenon that has for a long time appeared to have some connection to consciousness is *higher-order thought*. This is thought *about* your own thoughts; it involves taking a step back from the current flow of your experience and formulating a thought about it: "Why am I in such a bad mood?" or "I hardly noticed that car." Higher-order thought has long been seen as having a role to play in theories of subjectivity and consciousness, but it's been unclear what that role is. Some people have argued that higher-order thought is necessary for any sort of subjective experience at all. As most animals are very unlikely to have higher-order thought, the result is an extreme example of what I called a latecomer view of subjective experience. Another possibility is that higher-order thought is one of the sophisticated features of human life that has reshaped subjective experience in us, even though it did not bring experience into being.

I favor a view of that kind. I'd resist the idea that higher-order thought is *the* essential extra step that brings us to the kind of experience seen in humans. It's one piece of that story, though it may be an especially important part. Perhaps the most vivid of all forms of conscious thought are those in which we bring attention to bear on our own thought processes, reflect on them, and experience them *as* our own. We can look in at our own internal states without thinking in words about them, but in the undeniable *Why-did-I-think-that?* or *Why-do-I-feel-that-way?* cases of consciousness, inner speech is prominent. We often reflect on our inner states by forming inner questions, commentaries, and exhortations about them, and this is not idle or merely recreational; it can help us do things we'd not otherwise be able to do.

~ *Full Circle*

No one knows how old human language is—perhaps half a million years old, perhaps less—and there's much debate about how it evolved from simpler forms of communication. However language arose, its appearance changed the course of human evolution. By some path that we can presently only speculate about, language was also internalized; it became part of the machinery of thought. This internalization—Vygotsky's transition—was also an important evolutionary event. It is the second great internalization discussed in this book. The first, hundreds of millions of years earlier, was described in chapter 2. There, near the beginning of animal evolution, cells that had evolved means of sensing and signaling for use in interaction with each other, and with the rest of the external environment, gave these devices new roles. Cell-cell signaling was used to build multicellular animals, and within some of them a new control device arose: the nervous system.

The nervous system arose through one internalization of sensing and signaling, and the internalization of language as a tool for thinking was another. In both cases, a means of communication between organisms became a means of communication within them. These two events bookmark cognitive evolution as it has occurred to date—one near its outset and one in recent times. The recent one is not near the "end" of the process, but it is near the end of the process as it has run thus far.

In other respects these two internalizations have different shapes. In the evolution of the nervous system, the internalization of signaling was achieved by making the organism bigger—by expanding the boundaries of the organism to include formerly independent beings. In the internalization of language, the organism's boundaries remained unchanged, but a novel path within them was established.

In chapter 4 I looked at the evolutionary shift from a simple forward-directed flow linking the senses and action, to something more tangled. In the simplest cases, there's sensory input and some kind of output: what you do depends on what you see. Even in a bacterium, a causal arrow also runs the other way—an action has a *de facto* effect on what is sensed later. But in animals with nervous systems, the loops connecting sensing and acting become richer, and are registered *by* the animals themselves. Your actions continually change your relation to what's around you. That fact first appears as a *problem* for an animal trying to learn about the world. How can you track new events in your environment when everything you do changes how the world appears? But what begins as a problem can later become an opportunity.

In 1950, the German physiologists Erich von Holst and Horst Mittelstaedt introduced a framework for talking about these relationships. I used one of their terms earlier in this chapter: efference copy. I'll now outline some more of their framework. They used the term *afference* to refer to everything you take in through the senses. Some of what comes in is due to changes in the objects around you—that is *exafference* (with "ex" for outside, and pronounced in the same sort of way as "ex-boyfriend")—and some of what comes in is due to your own actions: that is *reafference* (pronounced "re-afference"). Animals face the challenge of distinguishing between the two. Reafference makes perception more ambiguous. If your own actions did not alter what your senses pick up, life would be easier in some ways.

One way to deal with the problem is with the "efference copy" mechanisms I described earlier. As you move, you send a signal to the parts of yourself that deal with perception, telling them to ignore some of what comes in: "Don't worry, that's just me."

Problems stem from reafference, but so do opportunities. You can affect your own senses in useful ways. Here the aim is not to filter out an unwanted contribution to what is perceived, but instead to use your actions to *feed* perception. A simple example is writing something down, a note to yourself, which you will read later. You act now, changing the environment, and later you will perceive the results of your act. This will enable you to do something, at that later time, that makes sense given what you know now.

To write a note and read it is to create a reafferent loop. Rather than wanting to perceive only the things that are *not* due to you—finding the exafferent among the noise of the senses—you want what you read to be *entirely* due to your previous action. You want the contents of the note to be due to your acts rather than someone else's meddling, or the natural decay of the notepad. You want the loop between present action and future perception to be firm. This enables you to create a form of external memory—as was, almost certainly, the role of much early writing (which is full of records of goods and transactions), and perhaps also the role of some early pictures, though that is much less clear.

When a written message is directed at others, it's ordinary communication. When you write something for yourself to read, there's usually an essential role for time—the goal is memory, in a broad sense. But memory of this kind *is* a communicative phenomenon; it is communication between your present self and a future self. Diaries and notes-to-self are embedded in a sender/receiver system, just like more standard kinds of communication.

Back in chapter 2 I also discussed two different roles that communication between individuals can have, roles that map onto different views of what the first nervous systems were doing for their owners. One role is to coordinate what is *perceived* with what is *done*; this is the role exemplified by Paul Revere's lantern code. The other role is to coordinate different components of a single action, such as when a person "calls the stroke" in a rowboat. In that earlier chapter I said that much of the time, both of these roles are being performed at once, but it's still worth distinguishing them. That's right, but we can also now see a connection between them that was not evident in that earlier discussion.

When you write something down in order to remind yourself to finish a job later, you are making a mark that your later self will *sense*—something you'll perceive. In that respect it's like the sexton and Revere. But that mark was made by your present self to get your later self to do something that completes a task. In that respect it's like the internal coordination of activities—action shaping—even though the coordination makes use of a causal loop that runs through the external world. The coordination involves making a mark that will later be sensed.

Some of these useful loops run outside the skin, and some run inside. Efference copies are internal messages, activity in the nervous system. When you move your head and the world seems to remain still, that is achieved by internal means. Here an internal message is used to solve a problem that arises from the effect of action on sensing. But these internal arcs, like the external ones, can also provide opportunities and novel resources. That is how things look within the model I gave earlier for the origins of inner speech. Copies of things you plan to say can give rise to silent actions of their own—inner actions that raise possibilities, put ideas together, and exert self-control. Inner speech can feel a bit like reafference—like the result of an action that affects your senses—but inner speech is confined inside, hence not really *heard* (at least when things are working as they should). If inner speech is a kind of broadcasting of information in the brain, it resembles the loop of reafference seen when you talk aloud to yourself or write notes to yourself. But this time the loop is tighter and more confined, invisible rather than public, a field for free and silent experiment.

When we see the human mind as the locus of countless loops of this kind, it gives us a different perspective on our own lives and those of other animals. This includes the cephalopods discussed in this book. Their expressive medium, colors and patterns, does not lend itself to complex loops. (That's true even setting aside the ironies associated with their alleged color-blindness.) Making skin patterns, no matter how complicated they might be, is more of a one-way street. The animal can't see its own patterns in the way a person can hear what they say. There's probably not much role for efference copies that involve skin patterns (unless some speculative theories of the role of chromatophores in skin sensing are correct). Cephalopod displays have enormous expressive power, but as long as we're looking at a single animal, rather than a pair or a group, these displays are not embedded in a lot of looping feedback, and perhaps could never be. The human case—an extreme case—suggests that the opportunities associated with reafference help to drive the evolution of a more complicated mind. Cephalopods are on a different road.

And this is not the only aspect of cephalopod life that circumscribes their possibilities.

7

EXPERIENCE COMPRESSED

Decline

I started watching cephalopods closely, following them around in the sea, around 2008: first the giant cuttlefish, and then octopuses, once I'd learned to see them (they'd been all around me, of course, the whole time). I also started reading about them, and one of the first things I learned came as a shock. Giant cuttlefish, these large and complicated animals, have very short lives: just one or two years. The same is true of octopuses; one or two years is a common lifespan. The largest, the giant Pacific octopus, can make it to about four years at the outside.

I could hardly believe it. I had assumed that the cuttlefish I'd been interacting with were old, had met humans often and worked out how they behaved, and had seen many seasons pass in their patch of ocean. I had assumed this partly because they *seemed* old; they had a worldly look. They also seemed too big to be so young, as they were often two or three feet long. I realized that year, though, that I'd come across these cuttlefish in the early part of the breeding season, and all the animals I'd been visiting would soon be dead.

Indeed, that is how it went. Toward the end of that southern winter the cuttlefish entered a sudden decline. It was visible over weeks, sometimes over days, when I was able to follow a single individual. They spontaneously began to fall apart. Soon some were missing arms and clumps of flesh. They began to lose their magical skin. At first I thought some of them were producing white patches as part of a display, but a closer look showed that the outer layer of skin, the living video screen, was instead falling off, leaving plain white flesh behind. Their eyes went cloudy. As this process reaches its end, the cuttlefish is unable to control its height in the water. Once the decline starts, it occurs very quickly. Their health seems to drop off a cliff.

Once I knew this stage was coming, interacting with these animals, especially the friendly ones, became poignant. Their time was so short. This discovery also made the puzzle of their large brains even more acute. What is the point of building a large nervous system if your life is over in a year or two? The machinery of intelligence is expensive, both to build and to run. The usefulness of learning, which large brains make possible, seems dependent on lifespan. What is the point of investing in a process of learning about the world if there is almost no time to put that information to use?

Cephalopods are evolution's only experiment in big brains outside of the vertebrates. Most mammals, birds, and fish live a lot longer than cephalopods. More accurately, the mammals and birds *can* live longer, if they don't get eaten or encounter some other mishap. This is especially true of larger species, like dogs and chimps, but there are monkeys the size of a mouse that can live for fifteen years, and hummingbirds that can live for over ten. Many cephalopods seem both too big and too smart to race through their lives in the way they do. What is all the brainpower doing if an octopus is dead less than two years after hatching from the egg?

Could there be something about the sea which imposes a short life? I quickly found out that is not the answer. A strange-looking rock-dwelling fish that inhabits the same patch of sea as my cephalopods is from a group that includes fish who live to two hundred years of age. Two hundred! This seemed extraordinarily unfair. A dull-looking fish lives for centuries while the cuttlefish, in their splendor, and the octopuses, in their curious intelligence, are dead before they are two?*

Another possibility was that something about the mollusk body plan, or something about cephalopods, makes a short life inevitable. I sometimes hear people say this, but it can't be the answer. Nautiluses, the elegant but psychologically unimpressive cephalopods who steer their shells like submarines around the Pacific, can live for more than twenty years. That is several drawn-out decades of life for what biologists, unflatteringly, have called a "smell-and-grope scavenger." These animals are relatives of octopuses and cuttlefish, and they are not rushing through their lives at all.

All this gave rise to a very different sense of what an octopus's or cuttlefish's life is like—rich in experience, but incredibly compressed. It also gave rise to more puzzlement about the brains that make that experience possible.

~ *Life and Death*

Why don't cephalopods live for a longer time? Why don't we *all* live for a longer time? On mountainsides in California and Nevada there are pine trees that were alive when Julius Caesar

---

*The cephalopods' situation is reminiscent of Ridley Scott's movie *Blade Runner*, in which a class of artificial but human-like "replicants" are programmed to die after only four years. (In the book by Philip K. Dick on which the film was based, *Do Androids Dream of Electric Sheep?*, their early deaths are due to breakdown.) *Blade Runner*'s replicants, unlike cephalopods, know their fate.

was wandering around Rome. Why do some organisms live for dozens, hundreds, or thousands of years while others, in the natural course of events, do not see even a single year pass? Death from accident or infectious disease is no puzzle; the puzzle is death from "old age." Why, after living for a time, do we fall apart? This question is always lurking as the birthdays pass, but the short lives of the cephalopods make it vivid. Why do we age?

We tend to think about this, intuitively, as a matter of bodies *wearing out.* Someone might say: we must wear out eventually, just as an automobile does. But the automobile analogy is not a good one. An automobile's original parts will indeed wear out, but an adult human is not operating with his or her original parts. We are made of cells that are continually taking in nutrients and dividing, replacing old parts with new ones. Even a cell that stays alive for a long time is continually turning over its material (most of it, anyway). If you keep replacing the parts of an automobile with new ones, there is no reason why it should ever stop running.

Here is another way to see the puzzle. Our bodies are collections of cells. These cells are stuck together and work in a coordinated way, but they are just cells. Most of the cells that make us up are continually dividing, creating two from one. Suppose that, for some reason, these dividing cells were bound to get "old," even though the cells actually present now have not been around very long. That is, suppose that even newly arrived cells show the age of their *lineage*, and this age is responsible for a body's decay. But if that is how things work, why do bacteria and other single-celled organisms still exist? The individual bacteria that are around now are the products of cell divisions that took place in the recent past, but their cell lineages are billions of years old.

Imagine taking a lot of bacteria of a particular kind—the

familiar *E. coli*, perhaps—and putting them together in a clump. When those cells divide, their offspring cells stay in the same clump. So the clump persists as cells come and go. If conditions were favorable, that clump might persist for millions of years. The clump would be a kind of "body"—a big collection of cells. There is no reason for it to wear out or break down just because it is old. Again, the parts present *now* are *not* old; they're brand-new cells. If that clump of cells can live forever, replacing and replenishing, then why not the clumps that are *our* bodies?

You might now say: it is the arrangement of our cells that makes us different from the bacteria. We are not just a clump. This arrangement can break down, even if the cells are always new. But why can't new cells remake the right arrangement? Cells can generate the right arrangement when a person is conceived, born, and develops from a baby to an adult. Why can't the arrangement needed to keep you alive be constantly *re*generated by the newly arriving cells?

Explanations in terms of "parts wearing out" are not enough to resolve the problem. Even if there is a version of this idea that does make sense, it fits poorly with many observations of lifespan in animals. If "wearing out" is the issue, then animals which have a faster metabolic rate—which burn through more energy—should age faster. This relationship has *some* predictive power, but it falls down in a fair number of cases. Marsupials such as kangaroos have lower metabolic rates than "placental" mammals like us, but they age faster. Bats have furiously active metabolisms, but they age slowly.

At the level of cells there is the possibility of indefinite renewal. But something about the kinds of objects we are—the kinds of collections of cells we are—gives us and other animals a relation to aging that is different from that of other living things. This way of looking at the issue takes us back many chapters, to the evolution of animals themselves. In animals, birth and death have come to exist as boundaries that mark out an individual life, even though cells are continually coming and going, and even though the cells' lineages extend before us and after us. So again we face the problem. Why do hummingbirds live till they are ten, rockfish till they are two hundred, bristlecone pines till they are thousands of years old, and octopuses until they are two?

~ A Swarm of Motorcycles

These puzzles have been largely resolved through some elegant pieces of evolutionary reasoning.

If we are thinking in evolutionary terms, it's natural to wonder if there is some hidden benefit from aging itself. Because the onset of aging in our lives can seem so "programmed," this is a tempting idea. Perhaps old individuals die off because this benefits the species as a whole, by saving resources for the young and vigorous? But this idea is question-begging as an explanation of aging; it assumes that the young *are* more vigorous. So far in the story, there's no reason why they should be.

In addition, a situation of this kind is not likely to be stable. Suppose we had a population in which the old do graciously "pass the baton" at some appropriate time, but an individual appeared who did *not* sacrifice himself in this way, and just kept going. This one seems likely to have the chance to have a few extra offspring. If his refusal to sacrifice was also passed on in reproduction, it would spread, and the practice of sacrifice would be undermined. So even if aging did benefit the species as a whole, that would not be enough to keep it around. This argument is not the end of the line for a "hidden benefit" view, but the modern evolutionary theory of aging takes a different approach.

The first step was made in the 1940s by a British immunologist, Peter Medawar, in a brief verbal argument. A decade later the American evolutionary biologist George Williams added a second step. A decade on again, in the 1960s, William Hamilton—probably *the* genius of late-twentieth-century evolutionary biology—put the new picture into rigorous mathematical form. Though the theory has been made precise in this way, the crucial ideas are satisfyingly simple.

Start with an imaginary case. Assume there is a species of animal with *no* natural decay over time. These animals show no "senescence," to use the word preferred by biologists. The animals start reproducing early in their life, and reproduction continues until the animal dies from some external cause—being eaten, famine, lightning strike. The risk of death from these events is assumed to be constant. In any given year, there is a (say) 5 percent chance of dying. This rate does not increase or decrease as you get older, but there is some number of years by which time *some* accident or other has almost certainly caught you. A newborn has less than a 1 percent chance of still being around at ninety years in this scenario, for example. But if that individual *does* make it to ninety, it will very probably make it to age ninety-one.

Next we need to look at biological mutations. Mutations are accidental changes to the structure of our genes. This is the raw material of evolution; very rarely, a mutation occurs which makes organisms better able to survive and reproduce. But the vast majority of mutations are harmful, or have no effect at all. Evolution produces what is called a *mutation-selection balance* with respect to many genes. This works as follows. Mutated forms of a gene are constantly entering the population, as a result of molecular accidents. Individuals with the mutated form are less likely to reproduce, so the bad mutations are eventually lost from the population. But even if every bad mutation is lost, that process takes time, and new mutations also keep coming in. So we expect a population to always contain some harmful mutated forms of each gene. A mutation-selection balance is a situation where bad mutations of a gene are being weeded out just as quickly as they are being introduced.

Mutations often tend to affect particular stages in life. Some act earlier, others act later. Suppose a harmful mutation arises in our imaginary population which affects its carriers only when they have been around for many years. The individuals carrying this mutation do fine, for a while. They reproduce and pass it on. Most of the individuals *carrying* the mutation are never *affected* by it, because some other cause of death gets them before the mutation has any effect. Only someone who lives for an unusually long time will encounter its bad effects.

Because we are assuming that individuals can reproduce through all their long lives, there is some tendency for natural selection to act against this late-acting mutation. Among individuals who live for a very long time, those without the mutation are likely to have more offspring than those who have it. But hardly anyone lives long enough for this fact to make a difference. So the "selection pressure" against a late-acting harmful mutation is very slight. When molecular accidents put mutations into the population, as described above, the late-acting mutations will be cleaned out less efficiently than early-acting ones.

As a result, the gene pool of the population will come to contain a lot of mutations that have harmful effects on long-lived individuals. These mutations will each become more common, or be lost, mostly through sheer chance, and that makes it likely that some will become common. Everyone will carry some of these mutations. Then if some lucky individual evades its predators and other natural dangers and lives for an unusually long time, it will eventually find things starting to go wrong in its body, as the effects of these mutations kick in. It will *appear* to have been "programmed to decline," because the effects of those lurking mutations will appear on a schedule. The population has begun to evolve aging.

The second main element in the theory was introduced by George Williams, an American biologist, in 1957. This is not a rival to the first idea; they are compatible. Williams's main point can be introduced by asking a simple question about saving for retirement. Is it worth saving enough money so that you will live in luxury when you are 120? Perhaps it is, if you have unlimited money coming in. Maybe you will live that long. But if you don't have unlimited money coming in, then all the money you save for a long retirement is money you can't do something else with now. Rather than saving the extra amount needed, it might make more sense to spend it, given that you are not likely to make it to 120 anyway.

The same principle applies to mutations. A lot of mutations have more than one effect, and in some cases, a mutation might have one effect that is visible early in life and another effect that is visible later. If both effects are bad, it is easy to see what will happen—the mutation should be weeded out because of the bad effect it has early in life. It is also easy to see what will happen if both effects are good. But what if a mutation has a good effect now and a bad effect later? If "later" is far enough away that you will probably not make it to that stage anyway, due to ordinary day-to-day risks, then the bad effect is unimportant. What matters is the good effect now. So mutations with good effects early in life and bad effects late in life will accumulate; natural selection will favor them. Once many of these have arisen in the population, and all or nearly all individuals carry some of them, a decay late in life will come to seem preprogrammed. Decay will appear in each individual as if on a schedule, though each individual will show the effects a bit differently. This happens not because of some hidden evolutionary benefit of the breakdown itself, but because the breakdown is the cost paid for earlier gains.

The Medawar effect and the Williams effect work together. Once each process gets started, it reinforces itself and also magnifies the other. There is "positive feedback," leading to more and more senescence. Once some mutations get established that lead to age-related decay, they make it even *less* likely that individuals will live past the age at which those mutations act. This means there is even less selection against mutations which have bad effects only at that advanced age. Once the wheel starts turning, it turns more and more quickly.

The picture I've been developing here is one full of pressures pushing downward on lifespan. But what about the thousand-year-old pine trees in California? They show no signs of falling apart. Trees are special, however, in two ways. First, they do not fit an assumption I made at an early stage of the argument above. I said that differences between how successful individuals are at reproducing late in life are not evolutionarily important because almost no individuals make it to that stage. But things are different if the few individuals who *do* succeed in reproducing when they are very old can have very large numbers of offspring. This is not true of us, but it is true of trees. Every branch on a tree is a site at which reproduction can take place, so a very old tree with its many branches can be much more fertile than a young tree. This enables trees to avoid some of the consequences of Medawar and Williams's arguments.

Second, a tree is a different kind of living thing from an animal, and to some extent the Medawar-Williams arguments do not even apply to it. The best way to approach this point is to first consider those "organisms" that can be seen, when you look closely, to actually be colonies. Some sea anemones, for example, form tight-knit colonies consisting of many small *polyps* which have a good degree of independence, especially in reproduction. One polyp can bud off another, and each polyp can make its own sex cells. These colonies can live indefinitely, in principle, just as a human society might, with individual humans coming and going but the society itself persisting.

Colonies and societies are not subject to the Medawar and Williams arguments, because they do not reproduce in the right way. The members of the colony or society (such as humans) can show senescence instead. An ordinary tree, like a pine or an oak, is not a colony, but it is not a single organism in quite the way a human being is, either. In some ways, it is in between those two cases. A tree grows by the multiplication of small units— branching stems—that can each reproduce on their own and, if cut and transplanted, can give rise to another tree. Anything that grows and develops through the multiplication of units that can reproduce in this way is exempt from the Medawar-Williams arguments.

I've introduced the two main ideas behind the evolutionary theory of aging. In the 1960s the theory was made rigorous and precise when the English evolutionary theorist William Hamilton turned his huge mind toward the problem. Hamilton recast the central ideas in mathematical form. Though this work tells us a good deal about why human lives take the course they do, Hamilton was a biologist whose great love was insects and their relatives, especially insects which make both our lives and an octopus's life seem rather humdrum. Hamilton found mites in which the females hang suspended in the air with their swollen bodies packed with newly hatched young, and the males in the brood search out and copulate with their sisters there inside the mother. He found tiny beetles in which the males produce and manhandle sperm cells longer than their whole bodies.

Hamilton died in 2000, after catching malaria on a trip to

Africa to investigate the origins of HIV. About a decade before his death, he wrote about how he would like his own burial to go. He wanted his body carried to the forests of Brazil and laid out to be eaten from the inside by an enormous winged *Coprophanaeus* beetle using his body to nurture its young, who would emerge from him and fly off.

> No worm for me nor sordid fly, I will buzz in the dusk like a huge bumble bee. I will be many, buzz even as a swarm of motorbikes, be borne, body by flying body out into the Brazilian wilderness beneath the stars, lofted under those beautiful and un-fused elytra [wing covers] which we will all hold over our backs. So finally I too will shine like a violet ground beetle under a stone.

~ Long and Short Lives

The evolutionary theory of aging gives us an explanation for the basic facts of age-related decay. It explains why breakdown starts to appear in old individuals as if on a schedule. More can be added to this outline to enable it to describe particular cases. In my thought-experiment above I assumed that reproduction takes place all through an organism's life. In many animals, including cephalopods, this is not even close to how things work.

Biologists distinguish between *semelparous* and *iteroparous* organisms. Semelparous organisms reproduce once, or in a single short season. This is also called "big bang" reproduction. Iteroparous organisms, like ourselves, reproduce many times over a more extended period. Female octopuses, in general, are an extreme case of semelparity—they die after a single pregnancy. A female octopus might mate with many males, but when it is time to lay eggs, she settles permanently into a den. There the female will lay her eggs, and fan and tend them as they develop. This one clutch can contain many thousands of eggs. The brooding might take a month, or several months, depending on the species and the conditions (things are slower in cold water). When the eggs hatch, the larvae drift off into the water. Soon afterward the female dies.

I am generalizing here. There's at least one exception among the octopuses, a rare species found in Panama by the same team that studied the squid signals discussed in chapter 5, Martin Moynihan and Arcadio Rodaniche. In their species, the females can reproduce over a longer period. No one knows why they are an exception.

Cuttlefish are a little different, but they still fall within the "big bang" category. They are active only in a single breeding season, but both sexes can engage in many matings and the females can produce many batches of eggs in that season. The females do not tend and protect the eggs, as octopuses do, but glue them to rocks of a suitable kind and leave them, moving off to mate and lay again. Then, as I described at the start of this chapter, they rapidly fall apart.

Why should an organism devote all its resources to one brood, or one breeding season? Much depends, again, on the risk of death by predation and other external causes—especially on how this risk changes over an animal's lifetime. Suppose in some animal the juvenile stage is risky, but once you get to be an adult, you can expect to live for a while without being eaten. Then it makes sense for adults to reproduce more than once. That applies to fish and many mammals. If, on the other hand, the adult life stage is very risky, it might make more sense to "go for broke" as soon as you get to a stage where you can breed.

Seasons also play a role. There might be a good season for laying eggs, or for hatching. That will determine a timetable within each year; perhaps it makes sense to mate in spring, or in winter. Then the question becomes: During how many years should you try to reproduce? Initially it might seem obvious that there is no harm in leaving it open, at least, that you will be around for another couple of years. You *might* make it through. Why fall apart in the meantime? But here the Williams argument returns, along with the need to think about these evolutionary questions by considering vast numbers of individuals and many generations. In the abstract, you would like to live and mate forever—at least from an evolutionary point of view. But who will leave more descendants, an organism which spends everything on one mating season, or a rival which spends less now in the hope of reproducing again later? If you spend less now to save something for later, that will do you no good if animals of your kind have little chance of *making* it to the next breeding season. In that case, it is better to put everything into one mating season, embracing all the options which give you an advantage now, even at the cost of breakdown once the season is done.

Evolution can give a species a vast lifespan or a tiny one. Within animals, the 200-year-old rockfish and the cuttlefish are extreme cases, and humans are intermediates. We and the rockfish both mature fairly slowly and reproduce over a number of years, but the rockfish goes on for longer. It is a spiny, venomous creature that no one tries to eat. The cuttlefish, in contrast, races to become large and fertile, mates in one season, and then falls to pieces.

The lifespans of different animals are set by their risks of death from external causes, by how quickly they can reach reproductive age, and other features of their lifestyle and environment. That is why we can live for about a century, a nondescript fish can live for twice as long, a pine tree's life can run from John the Baptist's to your own, and a giant cuttlefish—with its wild colors and friendly curiosity—arrives and is gone in a couple of summers.

In the light of all this, I think it is becoming clearer how cephalopods came to have their peculiar combination of features. Early cephalopods had protective external shells which they dragged along as they prowled the oceans. Then the shells were abandoned. This had several interlocking effects. First, it gave cephalopod bodies their outlandish, unbounded possibilities. The extreme case is the octopus, with almost no hard parts at all, and neurons spread through the body instead of bones. Back in chapter 3 I suggested that this open-endedness, this sea of behavioral possibility, was crucial to the evolution of their complex nervous systems. It's not that the loss of a shell alone created the evolutionary pressure leading to those nervous systems. Rather, a feedback system was established. The possibilities inherent in this body create an opportunity for the evolution of finer behavioral control. And once you have a larger nervous system, this makes it worthwhile to further expand the body's possibilities—collecting all those sensors on the arms, creating the machinery of color change and a skin that can see.

The loss of the shell also had another effect: it made the animals much more vulnerable to predators, especially fast-moving fish, with bones and teeth and good vision. That put a premium on the evolution of wiles and camouflage.

But there is only so much those tricks will achieve, only so many times they will save the animal. An octopus can't expect to live a long time, especially as they must be active as predators themselves. They can't just hide in a hole and wait for food to come to them. They have to be out and about, and once in the open they are vulnerable. This vulnerability makes them ideal candidates for the Medawar and Williams effects to compress their natural lifespan; a cephalopod's lifespan has been tuned by the continual risk of not making it to the next day. As a result, they have ended up with their unusual combination: a very large nervous system and a very short life. They have the large nervous system because of what those unbounded bodies make possible and the need to hunt while being hunted; their lives are short because their vulnerability tunes their lifespan. The initially paradoxical combination makes sense.

This picture is supported by the recent discovery of an exception to the usual cephalopod pattern, an exception that illuminates the rule. Most of what I've said about octopuses has been based on species that live in fairly shallow water, among reefs and shorelines. Much less is known about species that live in the true depths of the sea. A marine research unit at Monterey Bay, California (MBARI), explores deep-sea environments with remote-controlled submarines that carry video cameras. In 2007 they were inspecting a rocky outcrop nearly a mile underwater, off the coast of central California. They saw a deep-sea octopus (*Graneledone boreopacifica*) moving around. Returning a month or so later, they found the same octopus guarding a clutch of eggs. They kept returning to the site to watch the progress of this clutch of eggs, and always found the octopus there. In the end they watched this one octopus for four and a half years.

This octopus brooded her eggs for longer than any other known octopus is thought to live in total, and the fifty-three months it spent there is the longest egg-brooding period reported for any animal species. (For example, no fish is known to guard its eggs for more than four or five months.) It's not known how long this species of octopus can live, but as the report by Bruce Robison and his colleagues notes, if it spends the same fraction of its life brooding as other octopuses do, it might live for something like sixteen years.

This is strong evidence against any suggestion that octopus

bodies pose some physiological barrier to a long life. But why does this octopus live for so long when other species do not? The paper by Robison and his colleagues discusses how water temperature can make biological processes go more slowly. Deep waters are usually very cold (and I can't help being reminded of the fact that the one time I went scuba diving near Monterey I was colder than I've been in my life). Much of life runs in slow motion in cold water. Robison and his coauthors think this is part of the reason the mother can stay alive for so long, apparently without feeding. The paper also notes that the long brooding enables the young to hatch in a large and advanced state. Robison thinks that in this environment the egg's lengthy development gives an octopus a competitive advantage. I'd also suggest that the Medawar-Williams theory has a role to play, though. This theory predicts that predation risks should be much less severe for this species than they are for shallower-water octopuses, as the risk of predation affects an animal's "natural" lifespan. And here there is a strong clue. The MBARI images show an octopus sitting out in the open with her eggs for years on end. She did not find herself a den. Shallow-water octopuses do not, as far as I know, ever brood eggs out in the open like this. They would be sitting ducks for any predator that came along. In the deep sea, though, fish are much rarer than they are in the shallows. The fact that the Monterey octopus successfully brooded eggs in the open suggests that this species had less to fear from predation than other octopuses do. As a result, evolution has tuned its lifespan differently.

Putting these things together, we can see how many of the features of cephalopods—especially those features so pronounced in the octopus—might have stemmed from the abandoning of the shell all those years ago. This abandonment set them on a path of mobility, dexterity, and nervous complexity, and it also led to a live-fast-die-young lifestyle, an existence always exposed to the sharp-toothed predators around them.

~ Ghosts

I was diving in Sydney one day, a little away from my usual sites. Suddenly things went dark, and it was a moment before I realized I had swum into a massive cloud of ink. This was in an area scattered with boulders, gathered close with deep crevices between them. The inked area was the size of a large room. Everything was gunpowder gray, with thick black stringy shapes suspended here and there. There was too much ink to see what was going on, especially down in the crevices, and the ink hung there for a long time.

The next day I had a look in the same area. There was no ink, but I began to see dozens of cuttlefish eggs strewn across the sand at the bottom of some of the crevices. There was also a giant cuttlefish nearby. It was in awful condition. Its body was mostly white, and there was much damage on the arms. It watched me, hovering. Looking closer, I found three more, all quite large, clustered under a Stonehenge-like structure, with a natural rock roof, that rose several meters off the bottom of the sea. One cuttlefish was clearly male and the others seemed to be female. But it was hard to tell; they were all in various stages of decay. The worst off had lost much of their skin, leaving bare pearl-white bodies underneath, with fanning and crisscrossed cracks like broken glass in the skin that remained. Those who had more skin were pale gray. Some had eyes in very bad shape. A fifth cuttlefish, with some strong yellow left in her skin, swam in. But five of her arms were largely gone, and there were dark wounds in the flesh that remained. She swam off.

The four cuttlefish drifted about close to one another, wafting in tiny currents among the rocks. The eggs strewn on the sea floor were puzzling. Giant cuttlefish usually attach eggs to the roof of a ledge of some kind, where they hang down like tulip bulbs. I could not tell if these eggs had come adrift from where they were supposed to be, or had been laid where they were now. The ink I'd seen the previous day suggested that something might have gone wrong, but I had no idea what. The cuttlefish paid no attention to the eggs; they seemed to be just waiting. They also appeared to watch me, but with very little display, and I was not sure that all of them could still see me at all. Pale and quiet, they looked like cephalopod ghosts.

For some days there were cuttlefish there. There seemed to be arrivals and departures. The eggs remained down at the bottom of a crevice, lying in dim light with silt around. Finally I was there when one of the female cuttlefish reached the end. She was floating just outside the crevice when I arrived. A lot of her skin had been lost, with patches of orange-brown remaining. Two of her arms had gone completely, and one of her feeding tentacles hung motionless.

She was still swimming, with her fins moving gently. As I watched, I realized that we were both climbing a little in the water column, leaving the rocky crevice. Soon two fish took an interest in her. A pink fish began to circle but did not attack. A large leatherjacket was more of a problem. It came in, looked and circled, and then began a series of attacks, trying to bite pieces out of the front of the cuttlefish, even though the victim was several times larger than the attacker. I tried to keep the fish away, but it would not retreat far, and resumed its assaults whenever it could.

In response to the first attacks, the cuttlefish just flinched and waved her arms, with no effect at all. The fish kept coming. I realized that my attempts to defend the cuttlefish seemed to cause more panic in the cuttlefish than the fish's attacks. I was too big to be that close.

The leatherjacket came in again and bit harder, and this time the cuttlefish inked at it. The fish was not much deterred and approached again. Now the cuttlefish inked more profusely, and also began to spiral slowly. We continued to rise, passively, in the water. With the slow spiral and with gray-black ink pouring out of her funnel, the cuttlefish looked like a lumbering airplane on fire—an airplane which rose rather than fell to earth. Either because of the ink or the height we had now reached in the water, the fish abandoned its attacks. But this was all the cuttlefish could do. As she kept rising, the spirals stopped. She came up through the last meter of water and was suddenly floating on the surface, completely still. The surface of the water was a mess of small waves, which now sloshed the cuttlefish back and forth. I left her there.

The cuttlefish's death was a transition from swimming deep in her quiet world, through a slow spiraling ascent, to drifting on the noisy surface of ours.

8

OCTOPOLIS

An Armful of Octopuses

These days, the main place I watch octopuses is the site we call *Octopolis*, fifty feet below the surface off the east coast of Australia. As you swim down, on clear days the site is an Oz-like emerald green. On other days it's more like gray soup. I began visiting soon after Matt Lawrence discovered the site in 2009. The numbers at the site go up and down, but octopuses are always there. On the most intense days we'll count more than a dozen, roaming, grappling, or just sitting, all in and around an arena just a few yards across.

Reports of clumps of octopuses had cropped up from time to time before, but Octopolis was the first site that could be visited year after year, with several animals always present and often interacting. Sometimes a single octopus seems to have some command of the site, but this command is often partial, as there are too many individuals for one octopus to deal with at once. At first we thought this might be a harem-type situation, with one male and many females, but that turned out not to be right. Quite often there are multiple males present, though not too close to each other. It's hard to tell the sex of an octopus without interfering with it. In many species, the main difference is a groove under a male's third right arm, which is used in mating. That arm is stretched out toward the female, sometimes from close by, sometimes from cautious long range. If she accepts it, then a packet of sperm is passed along the underside of the arm. The females often then store the sperm for some time before fertilizing their eggs.

From the outset we have been determined to interfere with the octopuses as little as possible. We do interact with them, but only when they want to interact. We never pull octopuses from their dens, let alone turn them over and inspect their undersides. So the only way of reliably telling who is male and who is female is to watch how they behave, and see who engages in the telltale arm stretches of a male. In this way we're often able to work out the sexes of some individuals at the site, though other cases remain uncertain. This evidence is enough for us to be sure that multiple males and multiple females are often present.

Initially Matt Lawrence and I would just go down and watch them, and whenever we came to the surface we'd wonder what the octopuses would do once we were gone. For a while we could only speculate, but soon the small underwater GoPro video camera systems became available. We bought a couple of these, put them on tripods, and started leaving them down with the octopuses.

The first time we recovered these cameras and watched the footage we had no idea what we'd see. Footage of octopus behavior with no divers or submarines around had rarely been taken before. Would they behave completely differently when only a small camera was watching, and do something completely new? As far as we can tell, they behave fairly similarly whether

or not we're there, though there is a little more roaming and interaction when we're absent. This was disappointing in one way—no
secret group acrobatics—but reassuring in another, as it confirmed
that our presence doesn't much bother them.

Here is a typical shot from one of these videos, with three
octopuses roving over the shell bed. The far one in the center is
about to "jet" off somewhere, and the one on the right is also
moving under jet power.

Soon after this work began, a biologist working in Alaska,
David Scheel, got in touch with me. David did his training
studying lions in Africa. He spent weeks slowly following small
groups of lions, day and night, in a Land Rover, recording how
they roamed and hunted. He then switched animals and is now
an expert on the largest octopus species, the giant Pacific octopus. These can weigh 100 pounds or more, and David sometimes has to wrestle one to the surface in freezing Alaskan water
and get the animal into a boat for study at his lab. His is not one
of the labs that routinely takes the animals apart, and he has

done a lot of work tracking octopus movements by attaching small transmitters to their bodies and releasing them. David was keen to do some work on a different species (in warmer water). Soon he began making the trip down to Australia, and we had another person squashed into Matt's boat as we chugged out to Octopolis.

With David's help our thinking about the site became more systematic, and we spent more and more time measuring and counting. David is also much better than I am at imparting order to the mass of video data we collect. He has a knack for sifting through the multi-armed chaos to find patterns and ask questions that can actually be answered. In the southern summer of 2015, joined also by Stefan Linquist, we spent a couple of days parked near the site on a larger boat, trying to cover just about every daylight hour with our unmanned video cameras. It's never quite possible to do this. One enemy of the cameras is the octopuses themselves. Our pale head-like cameras on little tripods might look a bit like intruders of some kind—perhaps static, erect, three-legged cephalopods. The cameras are sometimes closely inspected and occasionally attacked as they film. Then the file we end up with has a lot of close-up footage of suckers and bites. On other occasions enormous stingrays come sweeping into the site and knock everything over.

In January 2015 the stars aligned; we were able to record a lot of video and could not have picked a better time to do it. We saw an unprecedented level of activity, and some of the behaviors we'd occasionally seen earlier resolved into patterns. A single octopus, a large male, seemed determined to control access to the site. He policed it continually throughout daylight hours. He chased off some octopuses, fighting intensely if they did not retreat (as shown in some of the color photos in the middle of this book). He tolerated others—we think these were females—and sometimes herded them into dens if they wandered away.

As an octopus roams over the shell bed, both the roamer and those in their dens will probe and sometimes lash at each other with their arms. We've seen a lot of arm-probing over the years at the site, and I'd always thought of it in pugilistic terms—in our first paper I described "boxing" as a frequent behavior. But Stefan Linquist (an amiable person) found himself thinking of many of these interactions as "high fives"—as arm-slaps that seemed to facilitate recognition between individuals, or at least a registration of basic roles on the site. Sometimes two octopuses would probe or whip their arms, and then settle back into a relaxed pose. Other times, the arm pokes would be followed by a fight. The photo below shows an octopus approaching from the right-hand side of the image, and as it comes in two others are stretching out their arms to probe or "high-five" the incomer.

All these behaviors are accompanied by continual color changes. Some of the color changes at the site seem quite unorganized, and fit the "chatter" hypothesis I outlined in chapter 5. Sometimes one of our unmanned cameras will film an octopus who seems, as far as I can tell, to be sitting quietly on his own, not interacting with another octopus or anything else, and he will run through a series of colors and patterns for no apparent reason. But other colors and patterns have more point to them. When an aggressive male is about to attack another octopus, he will often turn dark, rise out of the seabed, and stretch his arms out in a way that magnifies his apparent size. Sometimes he will raise his mantle, the entire rear part of his body, over his head like this:

We call this the "Nosferatu" pose, after the silent-film vampire of that name, with his dark cloak and threatening appearance. We had seen cases of this pose before, but the male we watched trying to control the site in 2015 used it often. He would bear down on another animal, who had to decide what to do. Sometimes the other one fled; sometimes he stood his ground, and a fight ensued. The Nosferatu male was not always larger than the other octopus, but he very rarely lost a fight (only once, in fact, did he lose on film).

David Scheel was interested in the colors octopuses took on during these interactions, and he went back through our old films, charting hundreds of encounters between an aggressor and a target. He noticed that the darkness of skin color is a reliable predictor of how aggressive an octopus will be—whether it will advance, whether it will stand its ground if another is coming. In contrast, several kinds of pale displays are produced when an octopus is not willing to fight. One of these is a bland pale gray, the other a stark blotchy pattern. That blotchy pattern is also seen when cephalopods of various kinds are threatened by predators; it's called a deimatic display, and its usual interpretation is that it's a last-ditch attempt to startle or confuse the foe. This raises the possibility that the deimatic display is something that octopuses produce involuntarily whenever a threat is bearing down at them, and is not a signal to the other octopus when we see it at our site. However, a deimatic display is sometimes produced at our site when an octopus is making its way back to a den under the watchful eye of a more aggressive individual. Then there is no question of flight, or any attempt to startle. So we think this display may have been pressed into service at Octopolis as something like a display of submission or nonaggression. Dark colors and the Nosferatu pose, on the other hand, seem to be displays that convey the seriousness of an aggressive move.

I commissioned an artist to do a drawing that shows these differences in pattern more clearly. In the picture on the next page, drawn from a video frame, the octopus on the left is bearing down, in a very dark pattern, on the octopus on the right. The one on the right, which is much paler and has just half of its body in the "deimatic" display, is beginning to flee.

~ Origins of Octopolis

Matt suspected this was an unusual place when he discovered it, but he did not realize quite how unusual it was. The most similar report was a controversial one from the tropical waters of Panama, nearly thirty years earlier.

In 1982 Martin Moynihan and Arcadio Rodaniche reported finding an unusual-looking and hitherto undescribed octopus with bright stripes, living in a group of several dozen animals and sometimes sharing dens. They reported this as part of the study of reef squid that I described in chapter 5, the study that claimed that squid have a "language" of colors and patterns on their skin. Moynihan and Rodaniche had no photos or video of the wild animals (underwater photography was a very different matter back in 1982), and there was not a lot of data that would be truly compelling to biologists. Moynihan and Rodaniche prepared a fuller description of the octopus for publication, but it was rejected. The whole topic of Panama's gregarious striped octopus was met with skepticism from other biologists for years, to Moynihan and Rodaniche's frustration.

It remained an enticing set of anecdotes until 2012, when the animal reappeared in the commercial aquarium trade. Some live specimens made their way to California, where they were kept by Richard Ross and Roy Caldwell of the Steinhart Aquarium. In captivity, some of the unusual behaviors reported by Moynihan and Rodaniche were confirmed, and more added. In the lab these animals will tolerate each other and share dens. Females mate and lay eggs over an extended period—as discussed in chapter 7, octopus females usually brood one clutch of eggs and then die. The paper by Caldwell, Ross, and colleagues does not contain field observations, but says that a company collecting sea life in Nicaragua knows of a single site where they aggregate. A field study is being prepared at the moment.

In the meantime, we have Octopolis, and it's a very unusual site. The more common pattern seen in octopuses is that an individual will make a den, live there for a short time, perhaps a few weeks, and then leave it to set up another. Males meet females to mate—often from a distance, via the outstretched arm—but do not hang around to help the female as she broods the eggs. In general, there's not thought to be much interaction at all between adult octopuses. Even *Octopus tetricus*, the species at our site, appears to be a lot less social when observed elsewhere.

So what happened at Octopolis? Some parts of what follows are speculative, but here is the story we've put together. Some time ago a single object was dropped onto the sandy sea bottom, probably from a boat. The object was made of metal, but it's now completely overgrown with marine life. This object is only a foot or so long and high, as it sits above the sea floor, but it's a valuable piece of real estate. The largest octopus on the site tends to live under it, and sometimes a few fish insist on living there, too, huddling alongside an octopus who pretends not to notice them. This object, we think, was enough to "seed" the site, in the same way a single object can seed the growth of a crystal.

We think that a first octopus, or a few of them, made a den at the found object, and began bringing scallops in to eat. The discarded shells accumulated, and soon began to change the physical properties of the site. The shells are discs a few inches in diameter. They are a much better den-building material than fine sand, and soon a few more dens could be built on the outskirts of the first den. Those octopuses brought in still more scallops to eat, leaving still more shells. A positive feedback process was under way: the more octopuses that lived there, the more shells were brought in, and the more dens could be built. This led to still more shells being brought in, and so on.

Another possibility is that the dropping of the metal object coincided with the dropping of a first load of shells. This might have happened back before 1984, when scallop dredging was banned in the bay, or around 1990, when scallop collection by divers was also banned. That load of shells would have given the site a bigger kick-start. But since then, it seems likely that most of the shells were brought in over the years by the octopuses. They have, by hunting and bringing food home, transformed the site where they live.

Why did this "seeding" have such large effects at one particular site? The general area where the metal object fell offers unlimited food for an octopus, as it's a scallop bed. The scallops live singly, or in little clumps. They are good food for an octopus. Despite the unlimited food, the area has very few good places for an octopus den. The sea floor is a fine sand that is hard to dig a stable hole in, and the predators in the area are numerous and deadly. We've seen dolphins and seals come zooming in to probe at the octopus dens. Several kinds of sharks live in the area. Huge "carpet sharks," broad bottom-dwelling animals which look like old bomber planes, sometimes come and lie on our site for long periods as the octopuses huddle in their dens. Some years ago, Matt took a disturbing video a little way from the site, when an octopus was caught out in the open by a school of aggressive small fish—leatherjackets. These look like piranhas and gather in hundreds. They've taken a nip at me a couple of times. We don't know why this one octopus was targeted, but after some cautious feints the fish attacked *en masse* and tore it to pieces. The octopus first tried to defend itself, then frantically to flee, zooming toward the surface, but it was dead within a couple of minutes. After that I began to wonder how octopuses could survive in the area at all. Those fish are around much of the time, and octopuses frequently leave their dens to collect food. My best guess is that an octopus can travel a certain distance from its den in safety, even with fish watching, because if the fish attack the octopus can be back in the den before damage is done. If the octopus goes outside that range, all bets are off. Very possibly the smaller octopuses have more to fear than larger ones, but there's not much an octopus can do against a hundred darting piranhas.

The leatherjackets prowl around, the seals zoom in, and the sharks cruise by and sit on the site. The most dramatic intruders of all are probably no direct threat to the octopuses: occasionally the light suddenly goes dark and an enormous black stingray comes sweeping in. These animals can be about as wide as a car, and they cruise in with great slow undulations of their wings. The octopuses duck. Our cameras, as I noted earlier, are usually scattered.

Octopolis, with its deep shell-lined dens, seems an island of safety in a dangerous area, and this probably explains the octopuses' consistent presence. But that raises a new question: Why don't the octopuses eat one another? At the site I have seen octopuses as tiny as a matchbox and others with an arm span over a meter across, with all sizes in between. Larger octopuses might not prey on each other, due to the risk of fights, but what protects the tiny ones? Many octopuses are cannibalistic, including some close relatives of our Octopolis species. Why not in this case? This, too, might be because of the abundance of local food that does not put up a fight: all those scallops.

Scallops, incidentally, do have eyes, with an unusual design that includes a mirror behind the retina. They can swim by flapping their shells. The first time I saw one move I was startled: swimming castanets! But these eyes and swimming skills are not nearly good enough to make a difference when octopuses are after them. They are helpless in that situation.

Recapping the story as we see it: the intrusion of a foreign object made a rare safe den. The first octopuses brought back scallops to eat and left the shells there. Soon the shells accumulated so much that *they* were the surface of the site. Eventually the shell debris made it possible to dig out stable dens in which others could live. The shell bed now extends so far that a newly dug den need not be very close to the main one. We still don't entirely understand what the shell bed is making possible. Some of the dens are quite deep, at least forty centimeters or so, and we are pretty sure that some octopuses spend time entirely covered by shells, invisible. Octopuses may reach out and interact with each other beneath the surface, perhaps mating. We see movements of the shells coming from below with no octopus visible. As more octopuses settle there, their environment comes to consist more and more of the shells themselves.

Our second paper about the site discussed this as a case of "ecosystem engineering"—the reshaping of an environment by the behavior of the animals who live there. As we realized when we worked on that paper, it's not only the octopuses who have been affected by all this. Many other species seem to have been attracted to the site. Schools of fish now hover above it and zoom back and forth. This has sometimes interfered with our video data. Squid hang out, signaling to each other. The enormous carpet sharks who lie on the site are probably not primarily there to eat the octopuses; we caught one on video doing a spectacular ambush lunge into a school of fish above it. Baby sharks of another species lie on the shell bed for part of the year. Small decorated stingrays called banjo rays sit on the site also, with hermit crabs crawling about on their bodies.

All of these creatures are present in much higher concentrations than are seen in areas just away from the site. The octopuses have built an "artificial reef" through their shell-collecting behaviors, and this seems to have led to an unusual social life developing, a life of high densities and continual interaction.

One way to interpret our Octopolis observations is to think that they show that octopuses, of this species and perhaps others, are generally more social than people realize. Their signaling behaviors—the color changes, the displays—do suggest this. A growing number of other studies push in the same direction: they suggest that octopuses are more engaged with each other than had once been thought. In 2011, a study of a species closely related to our Octopolitans reported that octopuses could recognize other individual octopuses. A more controversial study from 1992 suggested that octopuses can learn by watching each other behave. Another interpretation, applicable to at least some of what we've seen, is that this particular site is unusual. In conjunction with the overall intelligence of octopuses, an unusual context has led to unusual behaviors. The octopuses have had to work out how to manage their lives in this setting, and some of the resulting behaviors are opportunistic and novel. They have had to work out how to get along.

I suspect we're seeing a mix of new and old behaviors—some long-standing ones, and some that are improvised modifications that arose by individual adaptation to unusual circumstances.

Octopolis is a place where several elements that are usually missing from octopus life, and that are relevant to the evolution of brains and minds, are present. There's a lot of interaction and social navigation, and a lot of feedback between what is done and what is perceived. The octopuses face an unusually complicated environment, because an important part of that environment is other octopuses. There is also constant manipulation and reshaping of the shell bed. They throw debris around, and the shells and other materials that are thrown often hit other octopuses. This might be a mere den-cleaning behavior, but it's a behavior that has new consequences in the crowded setting, as these projectiles do seem to affect the behavior of the octopuses who are hit. We are trying at the moment to work out whether some of the throws are targeted.

All this takes place in the context of the usual short octopus lifespan, as far as we know. Octopus life is brief and there is no care of the young once they hatch. Suppose these octopuses live until they are about two years old. Since 2009, then, several generations have lived at the site. Many octopuses must have come and gone since we started visiting, and the animals continually remake the same complicated semi-sociality. We can imagine extra evolutionary steps that could be taken in a situation like this. Suppose the interactions became more complicated, the signaling more refined, the densities even higher. The life of each animal would become more caught up with the lives of others, and this would start to show in the ongoing evolution of their brains. We saw in chapter 7 that lifespan is tuned by lifestyle, especially the threat of predation. If octopuses of this species could reliably make it through more years without being eaten, there is no reason why they could not eventually evolve a longer life-span, too.

I am not saying that all this could happen at Octopolis—it could not. This is one small site, a tiny fraction of the range occupied by the species. When octopus eggs hatch, the baby octopuses drift away rather than staying where they were born. Each, if it survives, settles somewhere and starts to wander. So there's no reason to believe that the octopuses at the site today are the children or grandchildren of others who lived there before. One site and a few years mean nothing in evolutionary terms. Arrangements like these would have to persist on a large scale for thousands of years to have much effect. But the site gives a glimpse of one possible road in the ongoing evolution of the octopus.

~ Parallel Lines

As we near the end of the book, let's look back at the evolution of bodies and minds. The oldest and most encrusted landmarks were described in chapter 2: the ancient capacities for sensing and behaving, the evolution of animals from single-celled life, the first nervous systems. Then followed the evolution of the bilaterian body plan, the plan we share with bees and cephalopods. Soon after bilaterians appeared, there was a fork in the tree, with one side leading to vertebrates and another to a large range of invertebrate groups—insects, worms, mollusks.

The to and fro of sensing and acting is characteristic of all known organisms, including single-celled life. In the transition to the first animals with nervous systems, the machinery of external sensing and signaling was turned inward, enabling coordination within these new larger living units. Whatever nervous systems might have been doing initially, the transition from Ediacaran to Cambrian saw a new regime for animal behavior and the bodies that enable it. Organisms became entangled in each other's lives in new ways, especially as predator and prey. The tree continued to branch, a few brains expanded, and two experiments in very large nervous systems arose, one on the vertebrate side and one in the cephalopods.

With those outlines in place, I'll look at some features of the tree of life that take on new relevance when we revisit them now. These are parts of the tree that become visible when we zoom in on some branches that in the early chapters were viewed only from further away. Looking first at the vertebrate side, we find ourselves and other mammals. But mammals are not the only vertebrates to evolve high degrees of intelligence. Fish and reptiles can do surprising things, but the main example I have in mind is birds such as parrots and crows. Vertebrate brains are all "variations on a theme," with much in common, but the branchings are still quite deep. The common ancestor of birds and humans, a lizard-like animal, lived perhaps 320 million years ago, sometime before the age of the dinosaurs. From there, large brains arose along several independent paths within vertebrates. I said in chapter 3 that the history of large brains has the rough shape of a **Y**, with a vertebrate branch and a cephalopod branch, but this was quite a simplification. A closer look at the vertebrate side shows important internal branchings.

I covered the early evolution of the cephalopods in chapter 3 and wrote chapters on octopuses and cuttlefish. Both are cephalopods, but they are different in many ways. How did the history go on their side? Clearly there was a major branching in the evolution of cephalopods; how deep is it?

For a while it was believed, based on the fossil record, that the first appearance of the group of cephalopods that includes octopuses, cuttlefish, and squid (a group called the *coleoids*) occurred during the time of the dinosaurs, perhaps 170 million years ago. They diverged into their varied and familiar forms during the latter part of the dinosaurs' reign and afterward.

In a famous paper from 1972, Andrew Packard argued that the evolution of these cephalopods occurred in parallel with the evolution of certain kinds of fish. From about 170 million years ago, some fish started to evolve into a familiar "modern" form. Earlier cephalopods were the old predators of the sea. Fish evolved new forms that competed with them, and the cephalopods evolved in response. This included the evolution of their complex behavior.

The idea that modern cephalopods arose in one recent burst can be taken to support the view that large cephalopod nervous systems appeared in a kind of onetime evolutionary accident, followed by some later diversification. People have quite often taken a hypothesis of "accidental intelligence" in these animals seriously. Certainly there has been a temptation to think that octopuses, in particular, have "too much" brain for animals living such brief and asocial lives. Whether accident or not, the historical picture that Packard and others put in place encouraged the view that there was a single process: the evolution of large brains by *the cephalopods,* with minor variations arising afterward.

The historical picture then changed. Packard based his view on fossil evidence, which is always sketchy with soft-bodied animals. Later, evidence from genetics was brought in, and the picture that resulted was different. The new view has it that the most recent common ancestor of octopuses, cuttlefish, and squid lived not 170 but 270 million years ago. That is the point at which an evolutionary split led to an "octopod" group on one side, including octopuses and the deep-sea *Vampyromorpha*, and a "decapod" (ten-footed) group on the other, including squid and cuttlefish.

Backdating this split by 100 million years puts the divergence of cephalopods into a very different evolutionary scenario. The time of the split is now in the Permian period, before the dinosaurs. Life in the oceans was very different then. There still might have been competition between cephalopods and fish, but the earlier date makes it much more likely that cephalopods evolved complex nervous systems at least twice, once in the lineage leading to octopuses and once in the lineage leading to cuttlefish and squid.

You might reply: it could be that the common ancestor of all those cephalopods had *already* gained a lot of behavioral complexity, and was the smartest animal living in the Permian seas. The date of the divergence does allow this reply. But other new evidence pulls against it. In 2015 the first octopus genome was sequenced. From the genes, we can read off some new information about how nervous systems are *built* within the lifetime of each individual. Building a nervous system requires sticking cells together in precise ways. In us, a family of molecules called *protocadherins* are used to do this. The same family of molecules, it turns out, are used when octopus nervous systems are built.

That's interesting: similar tools are used in both cases. But another finding came along with this. Those molecules used in the building of nervous systems have diversified in squid as well as octopuses, and they seem to have done this *separately*, after the split between those two groups. Octopus evolution includes one expansion of this family of molecules, and squid evolution has its own separate expansion. So these brain-building molecules have diversified at least *three* times, not just once in the cephalopods and once in animals like us.

The significance of this depends on the extent to which cuttlefish and/or squid are genuinely smart animals. (For these purposes, we can treat cuttlefish and squid as one group.) We have less knowledge of cuttlefish cognition than we have for octopuses, and even less evidence about squid. But incoming evidence does suggest considerable brainpower in the enigmatic cuttlefish as well.

An example is a recent study of memory by Christelle Jozet-Alves and her group in Normandy, France, on a smaller species of cuttlefish than the giants of my earlier chapters. Memory in animals has several varieties. An important kind of memory in human experience is *episodic* memory—memory of particular events, as opposed to memory of facts or skills. (Your memory of your last birthday is an episodic memory; your memory of how to swim is a *procedural* memory, and your memory of the location of France is a *semantic* memory.) Jozet-Alves and her group based their cuttlefish experiment on a famous series of experiments that seem to show something like episodic memory in birds, and the team on this study included a leading bird researcher, Nicola Clayton. In both the bird and cuttlefish studies they talk of "episodic-*like*" memory, because episodic memory in humans has such a vivid element of subjective experience, and they don't know whether this is true of the other animals.

Episodic-like memory in these tests was taken to be memory of *where* and *when* a certain kind of food was available; it is "what-where-when" memory. The cuttlefish tests went like this. First the researchers worked out which of two foods (crab, shrimp) was preferred by each cuttlefish, and the cuttlefish were then placed in a situation where each food was associated with a different visual clue in the tank. Their more preferred food (which turned out to be shrimp) was replenished more slowly than the other one; if they ate shrimp, it took three hours before shrimp would be available in that spot again, while crab was replenished after one hour. The cuttlefish did learn that if they were released into the tank just one hour after their last meal of shrimp, there was no point in going to the shrimp location again, as there would be nothing there. After a one-hour delay they went to the crab location. If the delay was three hours, they went for the shrimp.

The existence of episodic-like memory in all these groups— mammals like us, birds, cuttlefish—is a striking example of what is almost certainly parallel evolution in these different lines. I don't know if anyone has tried a similar experiment with octopuses, and I don't know how they would do with the task. The Jozet-Alves study shows quite complex cognition on the decapod branch, in brains that evolved somewhat separately from those in octopuses. In other words, this is evidence of parallel evolution of intelligence *within* the cephalopods. This buttresses the view that it was no accident that complex nervous systems evolved in cephalopods. It's not something that happened once and was kept on, with variations, in a couple of different lines. Instead there was an expansion of the nervous system within the octopus line, and another one, in parallel, in the other cephalopods.

The relation between octopuses and cuttlefish is looking quite analogous to the relation between mammals and birds. In the vertebrate line, a split around 320 million years ago led to mammals and birds, each evolving large brains within somewhat different bodies. In cephalopods, octopuses and cuttlefish are both built on the molluscan plan, but the separation between them has something like the same historical depth, and there was parallel evolution of large brains here, also.

The tree could be represented like this:

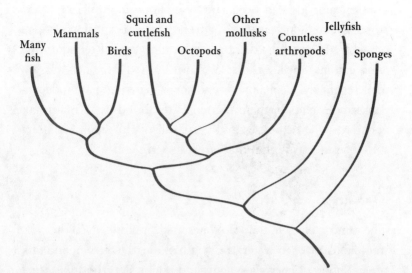

A Fragment of the Tree of Life: This drawing zooms in on some of the evolutionary branchings that figure in the book. The lengths of the "stems" between branch points are not to scale, and I've also represented groups of very different sizes on the same figure. Mammals and birds are large groups with respect to the number of species within them, while the two cephalopod groups on each side of their fork are much smaller. (Mammals and birds are each a *class*, in the traditional framework of biological classification, while all the cephalopods make up one class.) Arthropods, toward the right, are an entire *phylum*, consisting of a huge number of insects, crabs, spiders, centipedes, and others. Many groups are omitted from the diagram; if earthworms were included, for example, they'd be between the "other mollusks" and the arthropods, branching off that short stem leading to the mollusks. Starfish would be over near the vertebrates on the left. "Fish" don't make up a single branch. Most fish are on the far left branch, but some, such as the coelacanth, are on the branch that leads also to us and birds.

Cephalopods had been large predators since ancient times. About 270 million years or so ago, one group of cephalopods split, probably after they had embarked on their crucial relinquishing of an external shell. At least two lines evolved large nervous systems separately. Cephalopods and smart vertebrates are independent experiments in the evolution of the mind. Like mammals and birds, the octopuses and cuttlefish of this book represent sub-experiments within that larger experiment.

~ The Oceans

The mind evolved in the sea. Water made it possible. All the early stages took place in water: the origin of life, the birth of animals, the evolution of nervous systems and brains, and the appearance of the complex bodies that make brains worth having. The first ventures onto land probably took place not long after the history charted in my first chapters—certainly by 420 million years ago, perhaps earlier—but the early history of animals is a history of life in the sea. When animals did crawl onto dry land, they took the sea with them. All the basic activities of life occur in water-filled cells bounded by membranes, tiny containers whose insides are remnants of the sea. I said in chapter 1 that meeting an octopus is, in many ways, the closest we're likely to get to meeting an intelligent alien. Yet it's not really an alien; the Earth and its oceans made us both.

The features that made the sea productive of life and mind are invisible to us most of the time. They exist on a tiny scale. The sea does not change visibly as we do things to it—not in the way that cutting down a forest is immediately, undeniably visible. Waste poured into the sea just seems to drift and dilute away. As a result, the sea rarely appears urgent as an environmental problem, and measures we might take to help it often achieve little that's readily seen.

Sometimes the effects of our actions are visible once you take even a casual look below the surface. I began thinking about this book around 2008. I had bought a small apartment near the shore in Sydney, where I went during the northern summers. Like all the beaches up and down the coast around Sydney, this area had been fished for too long by too many people, and by the new millennium the waters had been just about emptied. But in 2002 one small bay was designated a marine sanctuary, with complete protection of its wildlife. Within a few years it was teeming with fish and other animals, and there I encountered the cephalopods that prompted me to write the book.

The efficacy of sanctuaries is encouraging, but the sea faces enormous threats. Overfishing the oceans is the most obvious, with more and more of what swims being hauled indiscriminately into the freezers of boats. Our ability to manage this is hampered not just by greed and competing interests, but by the difficulty of getting a handle on the problem and understanding our own destructive capacities. The sea looks the same after the boats are gone.

In the late nineteenth century, following the publication of *On the Origin of Species*, Thomas Huxley was Charles Darwin's most important scientific ally and a leading biologist in his own right. By the middle 1800s the fishers in the North Sea began to wonder whether they might exhaust their stocks of fish, and Huxley was invited to comment. He said there was little reason to worry. He did some simple calculations of the productivity of the sea and the fraction of fish being taken out, and concluded, in a speech in 1883: "I believe that it may be affirmed with confidence that, in relation to our present modes of fishing, a number of the most important sea fisheries, such as the cod fishery, the herring fishery, and the mackerel fishery, are inexhaustible."

He was spectacularly wrong in his optimism. Within a few decades many of these fisheries, especially the cod, were in serious trouble. As a result of his confident assurances, Huxley has become something of a villain. That is not unreasonable, though the villainizers do tend to overlook (and sometimes omit) a part of the infamous quote that I included above: "in relation to our present modes of fishing."

Huxley may have been wrong even with this qualification in place, but one thing that certainly sent people down the wrong path was their inability to recognize how much fishing technology would change. This led, in turn, to massive changes in how many fish each boat could take from the sea. With the growing mechanization of gear, then freezers, and high-tech means for tracking the fish, "our present modes of fishing" were gone not long after Huxley's optimistic words, and so were the fish.

Overfishing began in the nineteenth century and continues, with more meager returns, to this day. The other problem the sea faces is chemical change. This is even harder to see and more global in its sources, and, as a result, even harder to fix.

One example is acidification. As the CO_2 concentration in the atmosphere rises due to burning fossil fuels, some of the extra CO_2 dissolves into the sea. There it changes the water's pH balance, pushing it away from its usual state of mild alkilinity. The metabolisms of a great many sea animals, including cephalopods, are affected by this, and there are especially serious effects on corals and other organisms that make hard parts out of calcium. Those hard parts soften and dissolve in the altered sea.

In the later stages of writing this book I had lunch with a bee biologist, Andrew Barron. I met with him and Colin Klein, another philosopher, to discuss how we could possibly work out the evolutionary origins of subjective experience. When I heard that Andrew works on bees, I also wanted to ask him about "colony collapse," the problem that has been affecting bees worldwide.

This problem became apparent around 2007. Across many

countries, bee colonies quite suddenly started failing, and consequently failing to pollinate all sorts of crops that rely on them—
apples, strawberries, and many others. Given the economic
importance of bees as pollinators, the cause of the "collapses"
has been intensely studied. It has to be something worldwide,
not local. But the collapse came on pretty quickly. Is it a parasite?
A fungus? Chemical toxins? When I asked Barron, he said:
Yes, they are starting to get a handle on what's going on. So what's
the factor that's causing it? He replied that as far as they can tell
there is *no* single factor. Instead, over many years, more and more
small stresses have appeared in the lives of bees: more pollutants,
more new microorganisms, less habitat. For a long while, as these
stresses accumulated, bees were able to cope. Colonies absorbed the
stress by working harder. Although they weren't obviously and visibly suffering, the capacity of the bees to buffer these problems was
being slowly worn out. Eventually a critical point was reached, and
honeybee colonies just started to fail. They failed dramatically—
visibly—not because some sudden pest had swept through, but
because their capacity to absorb the stresses had run out. Now fruit
farmers desperately truck bee colonies thousands of miles from
orchard to orchard, trying to get their crops pollinated with the
bees that are still healthy enough to do the job.

I took that story on board, and now look at the ocean in the
same way. This sphere of biological creativity is so vast that for
centuries we could do whatever we liked to it and have little impact. But now our capacity to stress its systems is much greater. It
absorbs the stresses—not invisibly, but often in ways that are hard
to see, and easy to ignore when money is involved. In some places
it's already been pushed too far. In many parts of the world's seas
there are "dead zones," where no animals and little else can survive, due especially to the loss of oxygen. Dead zones probably
arose naturally from time to time before human stress on the

ocean, but they now occur on a much larger scale. Some of them come and go seasonally, following a malign rhythm set by fertilizer runoff from farms on land nearby, while others seem more permanent. "Dead zone": the very opposite of an ocean.

There are many reasons for us to appreciate and care for the oceans, and I hope this book has added one. When you dive into the sea, you are diving into the origin of us all.

NOTES

1. Meetings Across the Tree of Life

6 *The history of animals has the shape of a tree*: Darwin made extensive use of the "tree of life" idea in *On the Origin of Species*. Darwin was not the first to think about the relations between species as forming a tree shape, as he acknowledges. His innovation was to give the tree a historical, genealogical interpretation. In a sense, Darwin took the idea more *literally* than others had before him, in a way deftly expressed in a famous passage: "The affinities of all the beings of the same class have sometimes been represented by a great tree. I believe this simile largely speaks the truth." Charles Darwin, *On the Origin of Species by Means of Natural Selection, or the Preservation of Favoured Races in the Struggle for Life* (London: John Murray, 1859), 129.

For the history of tree thinking in biology, see Robert O'Hara, "Representations of the Natural System in the Nineteenth Century," *Biology and Philosophy* 6 (1991): 255–74. There are exceptions to the tree shape, especially outside the animals: see my *Philosophy of Biology* (Princeton, NJ: Princeton University Press, 2014). Richard Dawkins's book *The Ancestor's Tale: A Pilgrimage to the Dawn of Evolution* (New York: Houghton Mifflin, 2004) is a vivid and accessible description of the history of animal life, emphasizing the tree structure.

9 *This branch does not contain all the animals commonly known as "invertebrates"*: This term is regarded as problematic by some biologists, as it

does not pick out a definite branch of the tree, but organisms found on several branches. In this book I use various terms that some biologists disapprove of because they don't pick out definite branches of the tree; such terms include *prokaryote* and *fish*. I think these terms often remain useful.

11 *At the start of this book I placed an epigraph*: The first epigraph is from William James, *Principles of Psychology*, vol. I (New York: Henry Holt, 1890), 148. James, especially late in his career, was tempted by quite radical ways of achieving this "continuity" between the worlds of mind and matter—ways more radical than those pursued in this book. See "A World of Pure Experience," *The Journal of Philosophy, Psychology and Scientific Methods* 1, nos. 20–21 (1904): 533–43, 561–70.

12 *Does it feel like something to* be *one*: The phrase "all is dark inside" is taken from David Chalmers, *The Conscious Mind: In Search of a Fundamental Theory* (Oxford and New York: Oxford University Press, 1996), 96. Of course, all *is* dark inside a brain (outside of surgery). Things need not seem dark to the animal who possesses that brain, but the animal encounters the light by looking *out*side. In many ways the metaphor is quite misleading, but it does seem to capture something.

12 *The anthropologist Roland Dixon attributed to the Hawaiians*: The quote is from Roland Dixon, *Oceanic Mythology*, vol. 9 of *The Mythology of All Races*, ed. Louis Herbert Gray (Boston: Marshall Jones, 1916), 15. I am grateful to China Miéville, author of the cephalopodic novel *Kraken* (New York: Del Rey/Random House, 2010), for introducing me to Dixon and this passage.

2. A History of Animals

15 *The Earth is about 4.5 billion years old*: More exactly, the Earth *started* forming 4.567 billion years ago. For a treatment of the origin and early history of life, see John Maynard Smith and Eörs Szathmáry, *The Origins of Life: From the Birth of Life to the Origin of Language* (Oxford and New York: Oxford University Press, 1999). For a more technical presentation of some recent ideas, see Eugene Koonin and William Martin, "On the Origin of Genomes and Cells Within Inorganic Compartments," *Trends in Genetics* 21, no. 12 (2005): 647–54. Current views of the origin of life seem to be focusing on an origin within the sea itself, perhaps the deep sea, though other work has also looked at shallow pool-like environments. The date at which it is thought to be pretty clear that life existed is 3.49 billion years ago, so life evolved before that time. Life need not have begun with cells, but cells, too, are thought to be very old.

15 *Some of the early collaborations were probably so tight*: See Bettina Schirrmeister et al., "The Origin of Multicellularity in Cyanobacteria," *BMC Evolutionary Biology* 11 (2011): 45.

16 *Single-celled organisms can sense and react*: See Howard Berg, "Marvels of Bacterial Behavior," in *Proceedings of the American Philosophical Society* 150, no 3 (2006): 428–42; Pamela Lyon, "The Cognitive Cell: Bacterial Behavior Reconsidered," *Frontiers in Microbiology* 6 (2015): 264; Jeffry Stock and Sherry Zhang, "The Biochemistry of Memory," *Current Biology* 23, no. 17 (2013): R741–45.

17 *Those cells, eukaryotes, are larger and have an elaborate internal structure*: On the evolution of these more complex cells, and the role of an ancient swallowing of one cell by another, see John Archibald, *One Plus One Equals One: Symbiosis and the Evolution of Complex Life* (Oxford and New York: Oxford University Press, 2014). The swallower was only bacterium-like (as I say in the text) in an informal sense. It was probably an ancient archaean.

17 *Light, for living things, has a dual role*: For a general review, see Gáspár Jékely, "Evolution of Phototaxis," *Philosophical Transactions of the Royal Society B* 364 (2009): 2795–808. In 2016, a remarkable study described a cyanobacterium that may be able to focus an image by using the entire cell as a "microscopic eyeball," creating an image on the inside of the edge of the cell furthest from the source of light. See Nils Schuergers et al., "Cyanobacteria Use Micro-Optics to Sense Light Direction," *eLife* 5 (2016): e12620.

18 *They were also attracted to chemicals they could not eat*: See Melinda Baker, Peter Wolanin, and Jeffry Stock, "Signal Transduction in Bacterial Chemotaxis," *BioEssays* 28 (2005): 9–22.

18 *An example is* quorum sensing: See Spencer Nyholm and Margaret McFall-Ngai, "The Winnowing: Establishing the Squid-Vibrio Symbiosis," *Nature Reviews Microbiology* 2 (2004): 632–42.

19 *This aquatic setting is the right one*: For further discussion of this theme, see my "Mind, Matter, and Metabolism," *Journal of Philosophy*, in press.

19 *We're arriving at two thresholds*: For these relationships, very thoughtfully elaborated, see John Tyler Bonner, *First Signals: The Evolution of Multicellular Development* (Princeton, NJ: Princeton University Press, 2000). This book had a lot of impact on my thinking about behavioral transitions and multicellular life.

20 *Sensing and signaling between organisms gives rise to*: J. B. S. Haldane, one of the great evolutionists of an earlier generation, noted in 1954 that many hormones and neurotransmitters—substances used to control

and coordinate events within organisms like us—have effects on simple marine organisms when they encounter these chemicals in their environment. Chemicals we use as internal signals are interpreted by simpler organisms as external signals or cues. Haldane hypothesized that neurotransmitters and hormones have their origin in chemical signaling between some of our single-celled ancestors: see Haldane, "La Signalisation Animale," *Année Biologique* 58 (1954): 89–98. In the text I don't discuss hormonal systems that modify actions in real time, along with nervous systems. They are another interesting case of internal signaling.

20   *Animals are multicellular; we contain many cells that act in concert*: See John Maynard Smith and Eörs Szathmáry's classic *The Major Transitions in Evolution* (Oxford and New York: Oxford University Press, 1995), and a follow-up volume edited by Brett Calcott and Kim Sterelny, *The Major Transitions in Evolution Revisited* (Cambridge, MA: MIT Press, 2011). For a review of the many transitions to multicellularity seen in different groups, see Richard Grosberg and Richard Strathman, "The Evolution of Multicellularity: A Minor Major Transition?," *Annual Review of Ecology, Evolution, and Systematics* 38 (2007): 621–54. Even prokaryotes have evolved multicellular forms. I also discuss transitions to multicellularity in my *Darwinian Populations and Natural Selection* (Oxford University Press, 2009).

20   *The next stages in the history are unclear*: At the time of writing, this is an active controversy. A good presentation of what I call, in the text, the "majority" view is Claus Nielsen, "Six Major Steps in Animal Evolution: Are We Derived Sponge Larvae?" *Evolution and Development* 10, no. 2 (2008): 241–57. This view has been challenged by papers using genetic data to argue that ctenophores branched off from the rest of the animals before sponges did. See especially the paper by Joseph Ryan (and sixteen coauthors), "The Genome of the Ctenophore *Mnemiopsis leidyi* and Its Implications for Cell Type Evolution," *Science* 342 (2013): 1242592.

The fact that sponges (or ctenophores) are very distantly related to us does not mean that we had an ancestor who looked like a sponge (or ctenophore). A present-day sponge is the product of as much evolution as we are. Why should the ancestor look more like them than like us? But other factors come into play. If we look *within* the sponges, there are old evolutionary branchings that lead on both sides to a sponge-like sort of organism. It is possible, also, that sponges are *paraphyletic*—that they are not all descendants of a single common ancestor who branched off from other animals. If that's the case, it supports (though it definitely does not prove) the view that a sponge-like form was present in our past, because more than one lineage from that early time led to a sponge-like present-day animal.

For more on the hidden behaviors of sponges, see Sally Leys and Robert Meech, "Physiology of Coordination in Sponges," *Canadian Journal of Zoology* 84, no. 2 (2006): 288–306, and Leys's "Elements of a 'Nervous System' in Sponges," *Journal of Experimental Biology* 218 (2015): 581–91; Leys et al., "Spectral Sensitivity in a Sponge Larva," *Journal of Comparative Physiology* A 188 (2002): 199–202; and Onur Sakarya et al., "A Post-Synaptic Scaffold at the Origin of the Animal Kingdom," *PLoS ONE* 2, no. 6 (2007): e506.

23 *What nervous systems make possible*: In biology there are almost always exceptions: some neurons have direct electrical connections between them, and aren't restricted to using chemical signals to bridge the gap. Also, not all neurons have action potentials. For example, at the time of writing, it is unclear whether *Caenorhabditis elegans*, a tiny worm that is an important "model organism" in biology, uses action potentials at all in its nervous system. The system might work only with more smoothly graded and less "digital" changes in the electrical properties of its neurons.

For discussions of the evolution of neurons, see Leonid Moroz, "Convergent Evolution of Neural Systems in Ctenophores," *Journal of Experimental Biology* 218 (2015): 598–611; Michael Nickel, "Evolutionary Emergence of Synaptic Nervous Systems: What Can We Learn from the Non-Synaptic, Nerveless Porifera?" *Invertebrate Biology* 129, no. 1 (2010): 1–16; and Tomás Ryan and Seth Grant, "The Origin and Evolution of Synapses," *Nature Reviews Neuroscience* 10 (2009): 701–12. For a review of the ongoing debates, see Benjamin Liebeskind et al., "Complex Homology and the Evolution of Nervous Systems," *Trends in Ecology and Evolution* 31, no. 2 (2016): 127–35. Some biologists have argued that plants, too, have nervous systems. See Michael Pollan's "The Intelligent Plant," *New Yorker*, December 23, 2013: 93–105.

24 *As I see it, two pictures guide people's thinking*: For the history of this debate, and its significance, I am indebted to Fred Keijzer's work, and discussions with him.

Both the pictures I discuss here make the assumption that nervous systems are mostly for controlling *behavior*. This is a simplification, because nervous systems do many other things besides this. They control physiological processes such as sleep/wake cycles, and they guide large-scale changes in our bodies such as metamorphosis. Here, though,

I'll focus on behavior. The first tradition, emphasizing sensory-motor control, is a natural development of earlier philosophical ideas, but in explicit form it starts perhaps with George Parker's book *The Elementary Nervous System* (Philadelphia and London: J. B. Lippincott, 1919). George Mackie wrote some especially interesting papers in a framework continuous with Parker's—see Mackie's "The Elementary Nervous System Revisited," *American Zoologist* (now *Integrative and Comparative Biology*) 30, no. 4 (1990): 907–20, and Meech and Mackie, "Evolution of Excitability in Lower Metazoans," in *Invertebrate Neurobiology*, ed. Geoffrey North and Ralph Greenspan, 581–615 (Cold Spring Harbor, NY: Cold Spring Harbor Laboratory Press, 2007). This tradition is continued in Gáspár Jékely, "Origin and Early Evolution of Neural Circuits for the Control of Ciliary Locomotion," *Proceedings of the Royal Society B* 278 (2011): 914–22. Jékely, Keijzer, and I wrote a paper together that combines our ideas about the function of nervous systems and their early evolution; see Jékely, Keijzer, and Godfrey-Smith, "An Option Space for Early Neural Evolution," *Philosophical Transactions of the Royal Society B* 370 (2015): 20150181.

24 *This is* creating actions themselves: See Fred Keijzer, Marc van Duijn, and Pamela Lyon, "What Nervous Systems Do: Early Evolution, Input–Output, and the Skin Brain Thesis," *Adaptive Behavior* 21, no. 2 (2013): 67–85; and an interesting follow-up by Keijzer, "Moving and Sensing Without Input and Output: Early Nervous Systems and the Origins of the Animal Sensorimotor Organization," *Biology and Philosophy* 30, no. 3 (2015): 311–31.

25 *Above I treated interactions between neurons as a kind of signaling*: The important early model here is in David Lewis, *Convention: A Philosophical Study* (Cambridge, MA: Harvard University Press, 1969). His model was modernized by Brian Skyrms in *Signals: Evolution, Learning, and Information* (Oxford and New York: Oxford University Press, 2010). My "Sender-Receiver Systems Within and Between Organisms," *Philosophy of Science* 81, no. 5 (2014): 866–78, looks at how models of communication apply to interactions within the boundaries of one organism.

26 *Chris Pantin, an English biologist, developed the second view in the 1950s*: See C. F. Pantin, "The Origin of the Nervous System," *Pubblicazioni della Stazione Zoologica di Napoli* 28 (1956): 171–81; L. M. Passano, "Primitive Nervous Systems," *Proceedings of the National Academy of Sciences of the USA* 50, no. 2 (1963): 306–13; and the Fred Keijzer papers listed above.

27 *In 1946, an Australian geologist, Reginald Sprigg, was exploring some abandoned mines*: A biography of Sprigg called *Rock Star: The Story of*

Reg Sprigg—An Outback Legend was written by Kristin Weidenbach (Hindmarsh, South Australia: East Street Publications, 2008; Kindle ed., Adelaide, SA: MidnightSun Publications, 2014). Sprigg used his earnings as a geological explorer and entrepreneur to set up a sanctuary and ecotourism resort, Arkaroola. He also built his own deep-sea diving bell and held, at one time, a local scuba diving depth record (ninety meters, a depth at which you will never see me).

28 *I was shown around the exhibits by Jim Gehling*: The exhibit is at the South Australian Museum, Adelaide, where Gehling is senior research scientist. For my discussion of the Ediacaran, and the dates of various events in the history of animals, I have drawn extensively on Kevin Peterson et al. (including Gehling), "The Ediacaran Emergence of Bilaterians: Congruence Between the Genetic and the Geological Fossil Records," *Philosophical Transactions of the Royal Society B* 363 (2008): 1435–43. See also Shuhai Xiao and Marc Laflamme, "On the Eve of Animal Radiation: Phylogeny, Ecology and Evolution of the Ediacara Biota," *Trends in Ecology and Evolution* 24, no. 1 (2009): 31–40; and Adolf Seilacher, Dmitri Grazhdankin, and Anton Legouta, "Ediacaran Biota: The Dawn of Animal Life in the Shadow of Giant Protists," *Paleontological Research* 7, no. 1 (2003): 43–54.

30 *The clearest case is* Kimberella: This organism has had many interpretations, from jellyfish to mollusk. See M. Fedonkin, A. Simonetta, and A. Ivantsov, "New Data on Kimberella, the Vendian Mollusc-like Organism (White Sea Region, Russia): Palaeoecological and Evolutionary Implications," in *The Rise and Fall of the Ediacaran Biota*, ed. Patricia Vickers-Rich and Patricia Komarower (London: Geological Society, 2007), 157–79; and, more recently, Graham Budd, "Early Animal Evolution and the Origins of Nervous Systems," *Philosophical Transactions of the Royal Society B* 370 (2015): 20150037. For the molluscan interpretation, see Jakob Vinther, "The Origins of Molluscs," *Palaeontology* 58, Part 1 (2015): 19–34. During the period when this book was written, *Kimberella* became an ever more important and contentious fossil. One of my correspondents expressed concern that I was perpetuating a dubious interpretation of *Kimberella* as a mollusk; for another, *Kimberella*-as-mollusk is crucial to the interpretation of early bilaterian evolution. (These are not the authors of the papers cited above.) Perhaps by the time you read this, things will be clearer.

32 *Instead, in a phrase coined by the American paleontologist Mark McMenamin*: See Mark McMenamin, *The Garden of Ediacara: Discovering the First Complex Life* (New York: Columbia University Press, 1998).

33   *When the Royal Society of London held a conference*: The papers from this
     meeting have been published in *Philosophical Transactions of the Royal
     Society B* 370, December 2015. The meeting, titled Origin and Evolution
     of the Nervous System, was organized by Frank Hirth and Nicholas
     Strausfeld. For some of the discussion of jellyfish stings, see Doug
     Irwin's paper "Early Metazoan Life: Divergence, Environment and
     Ecology" in that collection. See also Graham Budd's paper "Early Ani-
     mal Evolution and the Origins of Nervous Systems." The next issue,
     volume 371, January 2016, has papers from a follow-up meeting, Ho-
     mology and Convergence in Nervous System Evolution, which was also
     of great value to this book.

33   *The "Cambrian explosion" began*: Here I use Charles Marshall, "Explain-
     ing the Cambrian 'Explosion' of Animals," *Annual Review of Earth and
     Planetary Sciences* 34 (2006): 355–84; and Roy Plotnick, Stephen Dorn-
     bos, and Junyuan Chen's "Information Landscapes and Sensory Ecol-
     ogy of the Cambrian Radiation," *Paleobiology* 36, no. 2 (2010): 303–17.

34   *The first bilaterians, or at least some early ones*: See Graham Budd and
     Sören Jensen, "The Origin of the Animals and a 'Savannah' Hypothe-
     sis for Early Bilaterian Evolution," *Biological Reviews*, published online
     November 20, 2015; and Linda Holland and six coauthors, "Evolution
     of Bilaterian Central Nervous Systems: A Single Origin?" *EvoDevo* 4
     (2013): 27. See also the *Philosophical Transactions of the Royal Society* vol-
     ume from the 2015 conference I discussed just above. Separate questions
     can be asked about the very *first* bilaterians and about the most recent
     common ancestor of all the bilaterians alive today. Eyespots, for example,
     may have been present in the latter but not the former. If the most recent
     common ancestor of living bilaterians had eyespots, that implies that
     Ediacaran bilaterian animals such as *Kimberella* and *Spriggina* had them
     (if they are bilaterian), or at least that their ancestors did. Again, it's all
     controversial at present.
          Starfish, by the way, are officially bilaterians, though in their adult
     form they have radial symmetry. There are controversies around the
     category; it has been argued that cnidarians are in fact bilaterian, or
     had a bilaterian ancestor. See John Finnerty, "The Origins of Axial
     Patterning in the Metazoa: How Old Is Bilateral Symmetry?," *Interna-
     tional Journal of Developmental Biology* 47 (2003): 523–29.

35   *The most behaviorally sophisticated animals outside the bilaterians*: See
     Anders Garm, Magnus Oskarsson, and Dan-Eric Nilsson, "Box Jelly-
     fish Use Terrestrial Visual Cues for Navigation," *Current Biology* 21,
     no. 9 (2011): 798–803.

36 *The first sophisticated eyes*: See Andrew Parker, *In the Blink of an Eye: How Vision Sparked the Big Bang of Evolution* (New York: Basic Books, 2003).

37 *As Budd sees it, animal behavior itself changed the way*: See Budd and Jensen, "The Origin of the Animals and a 'Savannah' Hypothesis . . . ," cited above. Gehling sketched hypotheses like this when he showed me around the Ediacarans in Adelaide.

38 *Michael Trestman, another philosopher, has offered*: See Trestman's paper "The Cambrian Explosion and the Origins of Embodied Cognition," *Biological Theory* 8, no. 1 (2013): 80–92.

39 *Here is one developed by the biologist Detlev Arendt*: See Maria Antonietta Tosches and Detlev Arendt, "The Bilaterian Forebrain: An Evolutionary Chimaera," *Current Opinion in Neurobiology* 23, no. 6 (2013): 1080–89; and Arendt, Tosches, and Heather Marlow, "From Nerve Net to Nerve Ring, Nerve Cord and Brain—Evolution of the Nervous System," *Nature Reviews Neuroscience* 17 (2016): 61–72.

41 *Here is a diagram of this part of the tree of life*: In this diagram I avoid taking sides on questions still in flux. Ctenophores are omitted altogether, though the indicated uncertainty about where neurons evolved reflects uncertainty about where ctenophores might be found on the tree. Starfish and other echinoderms, along with some other bilaterian invertebrate animals, are on our side of the fork. The diagram does not include organisms that are not animals, such as plants and fungi. These, and many single-celled organisms, would appear on branches further out to the right.

3. Mischief and Craft

43 *Claudius Aelianus*: The quote is from *On the Characteristics of Animals*, Book 13, translated by A. F. Schofield, Loeb Classical Library (Cambridge, MA: Heinemann, 1959), 87–88.

44 *Octopuses and other cephalopods are* mollusks: For the basics on the science of cephalopods and their behavior, see Roger Hanlon and John Messenger, *Cephalopod Behaviour* (Cambridge, U.K.: Cambridge University Press, 1996—a new edition may be out soon); and *Cephalopod Cognition*, a collection edited by Anne-Sophie Darmaillacq, Ludovic Dickel, and Jennifer Mather (Cambridge University Press, 2014). On the more popular side, see *Octopus: The Ocean's Intelligent Invertebrate*, by Mather, Roland Anderson, and James Wood (Portland, OR: Timber Press, 2010); and Sy Montgomery's book *The Soul of an Octopus: A Surprising*

Exploration into the Wonder of Consciousness (New York: Atria/Simon and Schuster, 2015).

45 *The cephalopod line probably goes back to an early mollusk*: For much of the history in this chapter I rely on Björn Kröger, Jakob Vinther, and Dirk Fuchs, "Cephalopod Origin and Evolution: A Congruent Picture Emerging from Fossils, Development and Molecules," *BioEssays* 33, no. 8 (2011): 602–13. James Valentine's book *On the Origin of Phyla* (Chicago: University of Chicago Press, 2004) gives the big picture.

45 *On dry land, no effortless move up into the air*: It is interesting that flight on land may have been invented, several times, in a more sea-like air. See Robert Dudley, "Atmospheric Oxygen, Giant Paleozoic Insects and the Evolution of Aerial Locomotor Performance," *Journal of Experimental Biology* 201 (1998): 1043–50.

46 *The nautilus, however, made it through*: For more on the nautilus, see Jennifer Basil and Robyn Crook, "Evolution of Behavioral and Neural Complexity: Learning and Memory in Chambered *Nautilus*," in *Cephalopod Cognition*, ed. Darmaillacq, Dickel, and Mather, 31–56.

47 *The oldest possible octopus fossil*: For the first, see Joanne Kluessendorf and Peter Doyle, "*Pohlsepia mazonensis*, an Early 'Octopus' from the Carboniferous of Illinois, USA," *Palaeontology* 43, no. 5 (2000): 919–26. Some biologists are not convinced by this one, which dates to more than 290 million years ago. The uncontroversial one, dated much later at around 164 million years ago, is called *Proteroctopus*. See J.-C. Fischer and Bernard Riou, "Le plus ancien octopode connu (Cephalopoda, Dibranchiata): *Proteroctopus ribeti* nov. gen., nov. sp., du Callovien de l'Ardèche (France)," *Comptes Rendus de l'Académie des Sciences de Paris* 295, no. 2 (1982): 277–80. The TONMO website has a good discussion of fossil octopuses: www.tonmo.com/pages/fossil-octopuses.

50 *As the cephalopod body evolved toward its present-day forms*: A good paper on this topic is Frank Grasso and Jennifer Basil, "The Evolution of Flexible Behavioral Repertoires in Cephalopod Molluscs," *Brain, Behavior and Evolution* 74, no. 3 (2009): 231–45.

50 *A common octopus* (Octopus vulgaris) *has about 500 million neurons*: Binyamin Hochner, in "Octopuses," *Current Biology* 18, no. 19 (2008): R897–98, gives this summary: "[T]he octopus nervous system contains about 500 million nerve cells, more than four orders of magnitude greater than in other molluscs (garden snails, for example, have around 10,000 neurons) and more than two orders of magnitude more than in advanced insects (cockroach and bee, for example, have around a million neurons), which probably rank next to cephalopods in invertebrate be-

havioral complexity. The number of neurons in the octopus is well into the range of amphibians such as the frog (~16 million) and small mammals such as the mouse (~50 million) and rat (~100 million), and not much fewer than in the dog (~600 million), cat (~1000 million) and rhesus monkey (~2000 million)."

It is difficult to count or estimate neurons, and these figures should be seen as rough. Suzana Herculano-Houzel of the Federal University of Rio de Janeiro has pioneered a new method and applied it to some animals, and octopuses are among the next on her list.

50 *The most startling finding in recent work on animal intelligence*: See Irene Maxine Pepperberg, *The Alex Studies: Cognitive and Communicative Abilities of Grey Parrots* (Cambridge, MA: Harvard University Press, 2000); Nathan Emery and Nicola Clayton, "The Mentality of Crows: Convergent Evolution of Intelligence in Corvids and Apes," *Science* 306 (2004): 1903–907; Alex Taylor, "Corvid Cognition," *WIREs Cognitive Science* 5, no. 3 (2014): 361–72.

51 *When biologists look at a bird, a mammal, even a fish, they are able to map*: See David Edelman, Bernard Baars, and Anil Seth, "Identifying Hallmarks of Consciousness in Non-Mammalian Species," *Consciousness and Cognition* 14, no. 1 (2005): 169–87.

52 *When tested in the lab, octopuses have done fairly well*: Hanlon and Messenger, *Cephalopod Behaviour*; *Cephalopod Cognition*, ed. Darmaillacq, Dickel, and Mather.

52 *Peter Dews was a Harvard scientist*: His paper is "Some Observations on an Operant in the Octopus," *Journal of the Experimental Analysis of Behavior* 2, no. 1 (1959): 57–63. For the history of thinking about learning through reward and punishment, see Edward Thorndike, "Animal Intelligence: An Experimental Study of the Associative Processes in Animals," *The Psychological Review*, Series of Monograph Supplements 2, no. 4 (1898): 1–109; and B. F. Skinner, *The Behavior of Organisms: An Experimental Analysis* (Oxford, U.K.: Appleton-Century, 1938).

55 *Octopuses in at least two aquariums have learned to turn off the lights*: One story is via the U.K. newspaper *The Telegraph*: The Sea Star Aquarium in Coburg, Germany, was troubled by mysterious blackouts. A spokesman said: "It was on the third night that we found out that the octopus Otto was responsible for the chaos . . . . We knew that he was bored as the aquarium is closed for winter, and at two feet, seven inches Otto had discovered he was big enough to swing onto the edge of his tank and shoot out the 2000-watt spotlight above him with a carefully directed jet of water" (www.telegraph.co.uk/news/newstopics

/howaboutthat/3328480/Otto-the-octopus-wrecks-havoc.html). Another case was at the University of Otago in New Zealand, described to me by Jean McKinnon (personal communication). She adds: "Doesn't happen anymore, we got waterproof lights!"

56 *Shelley Adamo, of Dalhousie University, had one cuttlefish*: Personal communication.

56 *In 2010, an experiment confirmed that giant Pacific octopuses*: See Roland Anderson, Jennifer Mather, Mathieu Monette, and Stephanie Zimsen, "Octopuses (*Enteroctopus dofleini*) Recognize Individual Humans," *Journal of Applied Animal Welfare Science* 13, no. 3 (2010): 261–72.

56 *Another tale that illustrates Linquist's point was told to me*: Jean Boal, personal communication.

59 *Many of these early experiments make for distressing reading*: Much of the early neurobiological work is like this—for example, various studies described in Marion Nixon and John Z. Young, *The Brains and Lives of Cephalopods* (Oxford and New York: Oxford University Press, 2003). The new EU rules are Directive 2010/63/EU of the European Parliament and Council.

59 *Jennifer Mather, along with Roland Anderson of the Seattle Aquarium, did the first studies of this behavior*: See Mather and Anderson, "Exploration, Play and Habituation in *Octopus dofleini*," *Journal of Comparative Psychology* 113, no. 3 (1999): 333–38; and Michael Kuba, Ruth Byrne, Daniela Meisel, and Jennifer Mather, "When Do Octopuses Play? Effects of Repeated Testing, Object Type, Age, and Food Deprivation on Object Play in *Octopus vulgaris*," *Journal of Comparative Psychology* 120, no. 3 (2006): 184–90. There is also a chapter in *Cephalopod Cognition* by the play expert Gordon Burghardt and Michael Kuba.

60 *The tour went on for ten minutes*: Matt timed it on his camera. This was not the only tour he's been led on by an octopus, though it was the longest.

60 *He posted some photos on a website*: The site is TONMO.com.

61 *The site we now call Octopolis*: Our first paper about the site is Godfrey-Smith and Lawrence, "Long-Term High-Density Occupation of a Site by *Octopus tetricus* and Possible Site Modification Due to Foraging Behavior," *Marine and Freshwater Behaviour and Physiology* 45, no. 4 (2012): 1–8.

62 *The next scene is on the shell bed itself*: This photo, and those on pages 101, 102, 181, and 182, are frames from videos taken by unmanned cameras at the site. Thanks to my collaborators Matt Lawrence, David Scheel, and Stefan Linquist for permission to print these in the book.

64 *In 2009, a group of researchers in Indonesia were surprised*: The paper is by Julian Finn, Tom Tregenza, and Mark Norman, "Defensive Tool Use in a Coconut-Carrying Octopus," *Current Biology* 19, no. 23 (2009): R1069–70. The best example of the use of compound tools by animals I know is the use by some chimpanzees of a stone anvil, for cracking nuts, together with a "wedge stone." The wedge stone is inserted under the anvil to level out its top surface, making for easier use. See William McGrew, "Chimpanzee Technology," *Science* 328 (2010): 579–80.

65 *In arthropods, very complex behaviors tend*: This is a broad generalization, and some writers would place much emphasis on the exceptions: spiders and stomatopods. For spiders, see Robert Jackson and Fiona Cross, "Spider Cognition," *Advances in Insect Physiology* 41 (2011): 115–74. Roy Caldwell, a leading octopus researcher at the University of California, Berkeley, claims that some stomatopods (or mantis shrimp) have very complex behavioral capacities and are not *less* sophisticated than octopuses, though because of their different sensory capacities, he thinks the comparison may not be very meaningful. See Thomas Cronin, Roy Caldwell, and Justin Marshall, "Learning in Stomatopod Crustaceans," *International Journal of Comparative Psychology* 19 (2006): 297–317.

65 *The ancestor at the center of the Y certainly had neurons*: There's an ongoing debate about the complexity of this animal, the protostome/deuterostome ancestor. See Nicholas Holland, "Nervous Systems and Scenarios for the Invertebrate-to-Vertebrate Transition," *Philosophical Transactions of the Royal Society B* 371, no. 1685 (2016): 20150047; and Gabriella Wolff and Nicholas Strausfeld, "Genealogical Correspondence of a Forebrain Centre Implies an Executive Brain in the Protostome-Deuterostome Bilaterian Ancestor," article 20150055 in the same issue of *Philosophical Transactions B*, which collects papers from the second day of the 2015 conference organized by Hirth and Strausfeld that I discussed in chapter 2.

My phrase "probably a worm-like creature" is supposed to be vague, not indicating a link to any particular kind of present-day worm (flatworms, annelids, etc.). Wolff and Strausfeld think, as their title says, that there was an "Executive Brain" in the common ancestor, but they have in mind a structure that is simple by most standards; they compare the hypothetical ancestor to flatworms with brains that contain hundreds of neurons. For a contrasting view, positing very small and simpler early bilaterians, see Gregory Wray, "Molecular Clocks and the Early Evolution of Metazoan Nervous Systems," article 20150046 in *Philosophical*

Transactions B 370, no. 1684 (2015), the collection of papers from the first day of that conference.

66 *On the other side, the cephalopods' side*: See Bernhard Budelmann, "The Cephalopod Nervous System: What Evolution Has Made of the Molluscan Design," in O. Breidbach and W. Kutsch, eds., *The Nervous System of Invertebrates: An Evolutionary and Comparative Approach*, 115–38 (Basel, Switzerland: Birkhäuser, 1995).

67 *Early work, looking at both behavior and anatomy*: See Nixon and Young, *The Brains and Lives of Cephalopods*.

67 *When an octopus pulls in a piece of food*: See Tamar Flash and Binyamin Hochner, "Motor Primitives in Vertebrates and Invertebrates," *Current Opinion in Neurobiology* 15, no. 6 (2005): 660–66.

67 *The nervous systems in each arm also include loops*: See Frank Grasso, "The Octopus with Two Brains: How Are Distributed and Central Representations Integrated in the Octopus Central Nervous System?" in *Cephalopod Cognition*, 94–122.

68 *described a very clever experiment*: See Tamar Gutnick, Ruth Byrne, Binyamin Hochner, and Michael Kuba, "*Octopus vulgaris* Uses Visual Information to Determine the Location of Its Arm," *Current Biology* 21, no. 6 (2011): 460–62.

In Sy Montgomery's book *The Soul of an Octopus*, she says that many researchers have anecodotes in which an octopus put into an unfamiliar tank with a piece of food seems to show disagreement across its arms. Some arms try to pull the animal toward the food, while others seem to want to cower in the corner. I once saw a situation that looked exactly like this, when an octopus was put into a tank at a lab in Sydney. The animal seemed to be pulled between arms that responded very differently to the situation. I am not confident about the significance of this event, though, especially as I realized later that the lights in the room were so bright that the animal may have been entirely confused.

69 *They rove around, often on reefs*: There are also deep-sea octopus species, about which less is known. There is a very good chapter about them in Darmaillacq et al.'s collection *Cephalopod Cognition*.

69 *When animal psychologists try to explain the evolution of a large brain*: See Nicholas Humphrey, "The Social Function of Intellect," in P. P. G. Bateson and R. Hinde, eds., *Growing Points in Ethology*, 303–17 (Cambridge, U.K.: Cambridge University Press, 1976); and Richard Byrne and Lucy Bates, "Sociality, Evolution and Cognition," *Current Biology* 17, no. 16 (2007): R714–23.

69 *To sharpen this idea up I'll adapt some ideas developed in the 1980s by the primatologist Katherine Gibson*: Her paper is "Cognition, Brain Size and the Extraction of Embedded Food Resources," in J. G. Else and P. C. Lee, eds., *Primate Ontogeny, Cognition and Social Behaviour*, 93–103 (Cambridge, U.K.: Cambridge University Press, 1986). I discuss these ideas also in "Cephalopods and the Evolution of the Mind," *Pacific Conservation Biology* 19, no. 1 (2013): 4–9.

71 *The demands of "social" life, in the within-species sense*: This point was made to me by both Michael Trestman and Jennifer Mather.

73 *Vertebrates and cephalopods separately evolved "camera" eyes*: See Russell Fernald, "Evolution of Eyes," *Current Opinion in Neurobiology* 10 (2000): 444–50; and Nadine Randel and Gáspár Jékely, "Phototaxis and the Origin of Visual Eyes," *Philosophical Transactions of the Royal Society B* 371 (2016): 20150042.

73 *Learning by attending to reward and punishment, by tracking what works*: See Clint Perry, Andrew Barron, and Ken Cheng, "Invertebrate Learning and Cognition: Relating Phenomena to Neural Substrate," *WIREs Cognitive Science* 4, no. 5 (2013): 561–82.

73 *Cuttlefish appear to have a form of* rapid eye movement *(REM) sleep*: See Marcos Frank, Robert Waldrop, Michelle Dumoulin, Sara Aton, and Jean Boal, "A Preliminary Analysis of Sleep-Like States in the Cuttlefish *Sepia officinalis*," *PLoS One* 7, no. 6 (2012): e38125.

74 *One central idea is that our body itself, rather than our brain*: A classic general discussion is Andy Clark's book *Being There: Putting Brain, Body, and World Together Again* (Cambridge, MA: MIT Press, 1997). For the robotics work, see Rodney Brooks, "New Approaches to Robotics," *Science* 253 (1991): 1227–32. The paper by Hillel Chiel and Randall Beer is "The Brain Has a Body: Adaptive Behavior Emerges from Interactions of Nervous System, Body and Environment," *Trends in Neurosciences* 23, no. 12 (1997): 553–57. Two interesting papers that make use of the concept of "embodiment" when thinking about octopuses are Letizia Zullo and Binyamin Hochner, "A New Perspective on the Organization of an Invertebrate Brain," *Communicative and Integrative Biology* 4, no. 1 (2011): 26–29, and Hochner's "How Nervous Systems Evolve in Relation to Their Embodiment: What We Can Learn from Octopuses and Other Molluscs," *Brain, Behavior and Evolution* 82, no. 1 (2013): 19–30.

The material at the end of this chapter was influenced by a discussion among a number of audience members at the Australasian Association of Philosophy Meetings in 2014, in response to a talk by Sidney

Diamante called "Reaching Out to the World: Octopuses and Embodied Cognition." Cecilia Laschi in Pisa currently heads a team working on a robotic octopus, with an emphasis on the arms: see www.octopus -project.eu/index.html.

75 *But that requires that there* be *a shape*: Technically, you might say that an octopus just has a *topology*—there are facts about which parts are connected to which, but the distances between parts and the angles are all adjustable.

75 *In an octopus, the nervous system as a whole is a more relevant object*: The optic lobes, behind the eyes, are sometimes described as not really part of the "central" brain, despite being important to octopus cognition.

4. From White Noise to Consciousness

78 *Thomas Nagel used the phrase* what it's like: See his "What Is It Like to Be a Bat?" *The Philosophical Review* 83, no. 4 (1974): 435–50.

78 *I don't claim to solve them entirely, but to take us closer*: Some additional steps are taken in "Mind, Matter, and Metabolism," forthcoming in *The Journal of Philosophy*, and in "Evolving Across the Explanatory Gap" (also forthcoming). Part of the solution will come from the development of new pieces of theory, and part will come from a critical reframing of the problem itself. I don't attempt much of that reframing here.

78 Subjective experience *is the most basic phenomenon that needs explaining*: I discuss some of these distinctions in more detail in "Animal Evolution and the Origins of Experience," in *How Biology Shapes Philosophy: New Foundations for Naturalism*, edited by David Livingstone Smith (Cambridge University Press, 2016).

79 *Nor is it something that pervades all of nature, as* panpsychists *believe*: See Thomas Nagel, "Panpsychism," in *Mortal Questions* (Cambridge, U.K.: Cambridge University Press, 1979), 181–95; and Galen Strawson et al., *Consciousness and Its Place in Nature: Does Physicalism Entail Panpsychism?*, ed. Anthony Freeman (Exeter, U.K., and Charlottesville, VA: Imprint Academic, 2006).

80 *Consider the case of* tactile vision substitution systems: See Paul Bach-y-Rita, "The Relationship Between Motor Processes and Cognition in Tactile Vision Substitution," in *Cognition and Motor Processes*, ed. Wolfgang Prinz and Andries Sanders, 149–60 (Berlin: Springer Verlag, 1984); also, Bach-y-Rita and Stephen Kercel, "Sensory Substitution and the Human-Machine Interface," *Trends in Cognitive Sciences* 7, no. 12

(2003): 541–46. For a more critical perspective on these technologies, see Ophelia Deroy and Malika Auvray, "Reading the World through the Skin and Ears: A New Perspective on Sensory Substitution," *Frontiers in Psychology* 3 (2012): 457.

81 *But their response was to reject the importance of input in a wholesale way*: I hope that sounds odd; how could you do *that*? Some philosophers put so much emphasis on the interpretation of experience by organisms that sensory "input" ends up being a sort of construction by the organism itself. Another approach, seen in biologically oriented philosophies more relevant to this book, is to expand the boundaries of the organism outward. Anything which plays a significant role in the to-and-fro of sensing and action must really be *internal* to the living system. A view of this kind has been defended recently by Evan Thompson, in his book *Mind in Life: Biology, Phenomenology, and the Sciences of Mind* (Cambridge, MA: Belknap Press of Harvard University Press, 2007). These views are often motivated by a determination to avoid the view that the organism is a passive recipient of information from outside. But they go too far the other way.

82 *The overall shape of the cause-effect relations looks like this*: See also Alva Noë, *Out of Our Heads: Why You Are Not Your Brain, and Other Lessons from the Biology of Consciousness* (New York: Hill and Wang, 2010), and Thompson, *Mind in Life*.

83 *Some fish, for example, send out electric pulses*: See Ann Kennedy et al., "A Temporal Basis for Predicting the Sensory Consequences of Motor Commands in an Electric Fish," *Nature Neuroscience* 17 (2014): 416–22.

83 *As the Swedish neuroscientist Björn Merker notes*: See his excellent paper "The Liabilities of Mobility: A Selection Pressure for the Transition to Consciousness in Animal Evolution," *Consciousness and Cognition* 14, no. 1 (2005): 89–114. Merker's paper had a good deal of influence on this chapter.

83 *This interaction between perception and action is also seen*: The importance of perceptual constancies to philosophical questions has been emphasized by Tyler Burge, in his *Origins of Objectivity* (Oxford and New York: Oxford University Press, 2010).

84 *But when this question was studied in pigeons*: See Laura Jiménez Ortega et al., "Limits of Intraocular and Interocular Transfer in Pigeons," *Behavioural Brain Research* 193, no. 1 (2008): 69–78.

85 *These experiments have also been done on octopuses*: See W. R. A. Muntz, "Interocular Transfer in Octopus: Bilaterality of the Engram," *Journal of Comparative and Physiological Psychology* 54, no. 2 (1961): 192–95.

85 *In more recent years, animal researchers such as Giorgio Vallortigara*: See

G. Vallortigara, L. Rogers, and A. Bisazza, "Possible Evolutionary Origins of Cognitive Brain Lateralization," *Brain Research Reviews* 30, no. 2 (1999): 164–75.

85 *These findings are reminiscent of experiments on "split brain" humans*: See Roger Sperry, "Brain Bisection and Mechanisms of Consciousness," in *Brain and Conscious Experience*, ed. John Eccles, 298–313 (Berlin: Springer-Verlag, 1964); Thomas Nagel, "Brain Bisection and the Unity of Consciousness," *Synthese* 22 (1971): 396–413; and Tim Bayne, *The Unity of Consciousness* (Oxford and New York: Oxford University Press, 2010).

86 *Marian Dawkins did a simple experiment*: Marian Dawkins, "What Are Birds Looking at? Head Movements and Eye Use in Chickens," *Animal Behaviour* 63, no. 5 (2002): 991–98.

87 *Evolution includes an awakening on a different*: There is also a third time-scale, that of individual development. See Alison Gopnik's *The Philosophical Baby: What Children's Minds Tell Us About Truth, Love, and the Meaning of Life* (New York: Farrar, Straus and Giroux, 2009).

88 *DF has been studied extensively by the vision scientists David Milner and Melvyn Goodale*: See (!) their book *Sight Unseen: An Exploration of Conscious and Unconscious Vision* (Oxford and New York: Oxford University Press, 2005). Here is the right place to mention an interesting criticism of some of the work I use in these passages, with respect to how it identifies "unconscious" processes. Does this work treat the presence of conscious experience too much as a yes-or-no matter? Perhaps it should instead be seen as entirely a matter of degree, in which case the collecting of data and reporting of results should be different. See Morten Overgaard et al., "Is Conscious Perception Gradual or Dichotomous? A Comparison of Report Methodologies During a Visual Task," *Consciousness and Cognition* 15 (2006): 700–708.

89 *In the 1960s, David Ingle rewired the nervous systems of some frogs*: His paper was "Two Visual Systems in the Frog," *Science* 181 (1973): 1053–55. The Milner and Goodale quote is from their book *Sight Unseen*.

90 *A view like this has been defended also by the neuroscientist Stanislas Dehaene*: See his *Consciousness and the Brain: Deciphering How the Brain Codes Our Thoughts* (New York: Viking Penguin, 2014). For more discussion of the eyeblink findings in the next paragraph, see Robert Clark et al., "Classical Conditioning, Awareness, and Brain Systems," *Trends in Cognitive Sciences* 6, no. 12 (2002): 524–31.

91 *Baars suggested that we are conscious of the information*: See Bernard Baars, *A Cognitive Theory of Consciousness* (Cambridge, U.K.: Cambridge University Press, 1988).

91–92   *My colleague at the City University of New York, Jesse Prinz*: See Jesse
Prinz, *The Conscious Brain: How Attention Engenders Experience* (Oxford
and New York: Oxford University Press, 2012).

92   *The result is what I'll call* latecomer *views about subjective experience*: See my
"Animal Evolution and the Origins of Experience" for more on this idea.

92   *But some of these people think there's no distinction*: Prinz takes this view.
I am not sure about Dehaene.

93   *Consider the intrusion of sudden pain*: Here I make use of some recent
works on pain in fish, birds, and invertebrates. The main ones are T.
Danbury et al., "Self-Selection of the Analgesic Drug Carprofen by
Lame Broiler Chickens," *Veterinary Record* 146, no. 11 (2000): 307–11;
Lynne Sneddon, "Pain Perception in Fish: Evidence and Implications
for the Use of Fish," *Journal of Consciousness Studies* 18, nos. 9–10 (2011):
209–29; C. H. Eisemann et al., "Do Insects Feel Pain?—A Biological
View," *Experientia* 40, no. 2 (1984): 164–67; R. W. Elwood, "Evidence for
Pain in Decapod Crustaceans," *Animal Welfare* 21, suppl. 2 (2012): 23–27.
For Derek Denton's work on "primordial emotions," see D. Denton et al.,
"The Role of Primordial Emotions in the Evolutionary Origin of Con-
sciousness," *Consciousness and Cognition* 18, no. 2 (2009): 500–514.

95   *The title of this chapter borrows a phrase from a paper by Simona Ginsburg
and Eva Jablonka*: The paper is "The Transition to Experiencing: I.
Limited Learning and Limited Experiencing," *Biological Theory* 2,
no. 3 (2007): 218–30.

97   *The Cambrian—with all its richer forms of engagement*: There are lots of
options here. It might be a mistake to see a *beginning* to subjective experience at this stage, as opposed to a change in degree and character. I
discuss some of the more radical options in "Mind, Matter, and Metabolism," forthcoming in *The Journal of Philosophy*.

97   *Then there were at least three separate origins*: Here I assume that the
protostome/deuterostome common ancestor was simple, and leading a
simple Ediacaran life. As I discussed above, some people think this animal was more complex, and had what Gabriella Wolff and Nicholas
Strausfeld call an "executive brain" that controlled choices of action: see
their "Genealogical Correspondence of a Forebrain Centre Implies an
Executive Brain in the Protostome-Deuterostome Bilaterian Ancestor,"
Philosophical Transactions of the Royal Society B 371 (2016): 20150055.
Their argument is based on similarities between the brains of present-
day vertebrates and arthropods (such as insects). Interestingly, they think
that cephalopods evolved a genuinely novel design, even if humans and
insects turn out to be refining the same ancestral plan: "[F]or cephalopod

molluscs, evidence overwhelmingly points to comparable behaviours driven by computational networks that have wholly independent ancestral origins." There's a question to ask here: the most recent octopus/human common ancestor is the same animal as the octopus/insect common ancestor. So it seems, on their view, that mollusks threw away their inherited "executive brain" and then cephalopods built a new one.

98 *Let's now return to the octopus*: Two groundbreaking papers on this question are Jennifer Mather, "Cephalopod Consciousness: Behavioural Evidence," *Consciousness and Cognition* 17, no. 1 (2008): 37–48, and Edelman, Baars, and Seth, "Identifying Hallmarks of Consciousness in Non-Mammalian Species," *Consciousness and Cognition* 14 (2005): 169–87.

99 *In an old 1956 experiment some octopuses were taught*: See B. B. Boycott and J. Z. Young, "Reactions to Shape in *Octopus vulgaris* Lamarck," *Proceedings of the Zoological Society of London* 126, no. 4 (1956): 491–547. Michael Kuba confirmed to me the surprising fact that there does not seem to have been any follow-up of this experiment, as far as he knows.

100 *Some years ago, Jennifer Mather did a careful study of this kind of behavior*: See her "Navigation by Spatial Memory and Use of Visual Landmarks in Octopuses," *Journal of Comparative Physiology A* 168, no. 4 (1991): 491–97.

102 *A recent paper written by Jean Alupay and her colleagues*: See Jean Alupay, Stavros Hadjisolomou, and Robyn Crook, "Arm Injury Produces Long-Term Behavioral and Neural Hypersensitivity in Octopus," *Neuroscience Letters* 558 (2013): 137–42, and also Mather, "Do Cephalopods Have Pain and Suffering?" in *Animal Suffering: From Science to Law*, eds. Thierry Auffret van der Kemp and Martine Lachance (Toronto: Carswell, 2013).

The study above by Alupay and her colleagues also found that when the parts of the octopus's central brain that are usually seen as the "smartest" were removed (the vertical and frontal lobes), this did not prevent the octopuses from doing their wound-directed behaviors. So, as the researchers say, either wound-directed behavior is not an indicator of pain in the way it is usually taken to be, or the octopuses have pain-related representations of their body somewhere else in their nervous system. I suspect the latter, though no one really knows.

103 *Let's consider some analogies with our case*: I'm grateful to Laura Franklin-Hall for making a number of interesting suggestions on this point, during a discussion that followed a visit to Benny Hochner's octopus lab in Jerusalem.

104 *They are usually invisible to us, but they are there*: See M. A. Goodale, D. Pelisson, and C. Prablanc, "Large Adjustments in Visually Guided

Reaching Do Not Depend on Vision of the Hand or Perception of Target Displacement," *Nature* 320 (1986): 748–50.

105 *In the paper on "embodied cognition" I quoted earlier*: Chiel and Beer, "The Brain Has a Body: Adaptive Behavior Emerges from Interactions of Nervous System, Body and Environment," *Trends in Neurosciences* 23 (1997): 553–57.

5. Making Colors

109 *Alexandra Schnell, one of the few people*: See Alexandra Schnell, Carolynn Smith, Roger Hanlon, and Robert Harcourt, "Giant Australian Cuttlefish Use Mutual Assessment to Resolve Male-Male Contests," *Animal Behavior* 107 (2015): 31–40.

109 *Here is how it works*: Hanlon and Messenger's book *Cephalopod Behavior* has a good description. Many papers from Roger Hanlon's lab at the Woods Hole Marine Biological Laboratory follow up: www.mbl.edu /bell/current-faculty/hanlon. For detail on the chromatophores, see Leila Deravi et al., "The Structure-Function Relationships of a Natural Nanoscale Photonic Device in Cuttlefish Chromatophores," *Journal of the Royal Society Interface* 11, no. 93 (2014): 201130942. My sketch of the skin layers is loosely based on a figure in this paper. Not all cephalopods have the full three-layer machinery pictured here.

119 *This impossible conclusion is based*: See Hanlon and Messenger's *Cephalopod Behaviour*, Box 2.1, p. 19.

120 *The first pieces were put down in 2010*: See Lydia Mäthger, Steven Roberts, and Roger Hanlon, "Evidence for Distributed Light Sensing in the Skin of Cuttlefish, *Sepia officinalis*," *Biology Letters* 6, no. 5 (2010): 20100223.

120 *First, it's possible that these molecules*: All that the first paper established was that the *genes* for these molecules were active in the skin.

120 *I'd just sent off a book review*: Review of *Cephalopod Cognition*, ed. Darmaillacq, Dickel, and Mather, *Animal Behavior* 106 (2015): 145–47.

121 *The paper, written with Todd Oakley, showed first*: M. Desmond Ramirez and Todd Oakley, "Eye-Independent, Light-Activated Chromatophore Expansion (LACE) and Expression of Phototransduction Genes in the Skin of *Octopus bimaculoides*," *Journal of Experimental Biology* 218 (2015): 1513–20.

122 *Another possibility was suggested to me by Lou Jost*: This is on my old cephalopod website, http://giantcuttlefish.com/?p=2274.

122 *As chromatophores of different colors expanded and contracted*: Using this mechanism, if expanding a red chromatophore affected incoming light

less than expanding a yellow one does, that would show that the light contained more red.

124 *Yet the cuttlefish escaped*: Cephalopod ink contains more than dark coloring. It has compounds that may have various effects on the nervous systems of predators. See Nixon and Young, *The Brains and Lives of Cephalopods* (New York: Oxford University Press, 2003), 288.

124 *The original function of cephalopod color change*: The relation between camouflage and signaling functions is discussed in detail in Jennifer Mather, "Cephalopod Skin Displays: From Concealment to Communication," in *Evolution of Communication Systems: A Comparative Approach*, ed. D. Kimbrough Oller and Ulrike Griebel, 193–214 (Cambridge, MA: MIT Press, 2004).

125 *This is seen most dramatically in one place*: See Karina Hall and Roger Hanlon, "Principal Features of the Mating System of a Large Spawning Aggregation of the Giant Australian Cuttlefish *Sepia apama* (Mollusca: Cephalopoda)," *Marine Biology* 140, no. 3 (2002): 533–45. Some complex behaviors are seen here. Some males who are not large enough to act as consorts to females try to "impersonate" females, in order to escape a guarding male's vigilance and get close to females. They quite often succeed.

127 *Another possibility is connected to the speculative ideas*: This suggestion was made by Jane Sheldon.

128 *Wild baboons in the Okavango Delta of Botswana, Africa, have been studied for years*: Dorothy Cheney and Robert Seyfarth, *Baboon Metaphysics: The Evolution of a Social Mind* (Chicago: University of Chicago Press, 2007). See my "Primates, Cephalopods, and the Evolution of Communication," a paper to appear in a new collection about Cheney and Seyfarth's work, for more on their view. Baboons do also have a range of communicative gestures, as well as their calls.

Jennifer Mather's paper "Cephalopod Skin Displays: From Concealment to Communication" also discusses the unusual sender-receiver relationships in cephalopod displays.

130 *Signal production in one cephalopod species, the Caribbean reef squid, was documented*: The interesting work discussed here is Martin Moynihan and Arcadio Rodaniche, "The Behavior and Natural History of the Caribbean Reef Squid (*Sepioteuthis sepioidea*). With a Consideration of Social, Signal and Defensive Patterns for Difficult and Dangerous Environments," *Advances in Ethology* 25 (1982): 1–151. Arcadio Rodaniche passed away as this book was being completed. I am grateful to Denice Rodaniche for assistance with the history of Moynihan and Rodaniche's work.

132 *These squid are among the most social*: The aggregation of giant cuttlefish

in Whyalla is another case, though it's temporary—they come together to reproduce. Humboldt squid live in large aggregations. They have not been studied very much, in part because they are large and can be aggressive. They are perhaps the most aggressive of known cephalopods. Some recent observations of nautiluses by Julian Finn have found them in large groups, also.

6. Our Minds and Others

137 *In one of the most famous passages in all of philosophy*: The passage appears in David Hume's *A Treatise of Human Nature*, Book I, Part IV, Section VI, "Of Personal Identity," first published in 1739.

138 *Perhaps Hume was one of those for whom inner speech is weak*: Christopher Heavey and Russell Hurlburt found that inner speech takes up about 26 percent of the conscious waking life of a sample of college students. They also found a lot of variation across subjects. See Christopher Heavey and Russell Hurlburt, "The Phenomena of Inner Experience," *Consciousness and Cognition* 17, no. 3 (2008): 798–810.

139 *Nearly two centuries after Hume, the American philosopher John Dewey*: He made the comment in chapter 5 of his book *Experience and Nature* (Chicago: Open Court Publishing, 1925).

139 *Lev Vygotsky grew up in what is now Belarus*: Vygotsky's *Thought and Language* was published posthumously in 1934, the year of his death. It appeared in English in 1962, translated by Eugenia Hanfmann and Gertrude Vakar, and issued by the MIT Press. A revised and expanded edition of that translation followed in 1986, edited by Alex Kozulin, restoring Vygotsky's original text.

140 *A few prominent people working today, such as Michael Tomasello*: The (rightly) famous Tomasello book was *The Cultural Origins of Human Cognition* (Cambridge, MA: Harvard University Press, 1999). Andy Clark gives a lot of credit to Vygotsky in his pathbreaking book *Being There: Putting Brain, Body, and World Together Again* (Cambridge: MIT Press, 1997).

141 *Nicola Clayton and others at the University of Cambridge*: A couple of examples are Joanna Dally, Nathan Emery, and Nicola Clayton, "Food-Caching Western Scrub-Jays Keep Track of Who Was Watching When," *Science* 312 (2006): 1662–65; and Clayton and Anthony Dickinson, "Episodic-like Memory During Cache Recovery by Scrub Jays," *Nature* 395 (2001): 272–74.

142 *Köhler was a German psychologist who spent four years*: See his book *The Mentality of Apes*, trans. Ella Winter (New York: Harcourt Brace, 1925).

142 *Second, he used the remarkable case of the French Canadian monk*: Merlin
     Donald's book *Origins of the Modern Mind: Three Stages in the Evolution
     of Culture and Cognition* (Cambridge, MA: Harvard University Press,
     1991) is still very interesting, though it's an old book now. The "Brother
     John" paper is André Roch Lecours and Yves Joanette, "Linguistic and
     Other Psychological Aspects of Paroxysmal Aphasia," *Brain and Lan-
     guage* 10, no. 1 (1980): 1–23. In the text I use the past tense about Brother
     John, but I've been unable to find out if he's still alive.

143 *Extreme views on both sides of the question are fading*: Peter Carruthers,
     "The Cognitive Functions of Language," *Behavioral and Brain Sciences*
     25, no. 6 (2002): 657–74, is a good survey, and is followed by a set of
     comments by other researchers expressing alternative views.

143 *Here is an example from recent research on young children*: The study, pres-
     ently under review, is Shilpa Mody and Susan Carey, "Evidence for the
     Emergence of Logical Reasoning by the Disjunctive Syllogism in Early
     Childhood." They found that children younger than three did not suc-
     ceed at a task that requires processing a disjunctive syllogism, but three-
     year-olds did. They also note (citing other work) that although children
     use the word "and" shortly after their second birthday, they don't pro-
     duce "or" until they are about three. Mody and Carey are cautious about
     the interpretation of this finding, and they don't claim it shows that in-
     ternalization of this part of public language is enabling children to
     succeed on the task.
        A well-known experiment that pushes in a similar direction was
     done by Linda Hermer and Elizabeth Spelke: "A Geometric Process for
     Spatial Reorientation in Young Children," *Nature* 370 (1994): 57–59, with
     follow-up work and conclusions discussed in Spelke's "What Makes Us
     Smart: Core Knowledge and Natural Language," in Dedre Gentner and
     Susan Goldin-Meadow's collection, *Language in Mind: Advances in the
     Investigation of Language and Thought* (Cambridge, MA: MIT Press,
     2003). This work suggested that only humans who can use language are
     able to combine information of different kinds (geometry plus color cues)
     when trying to navigate a room, while rats and prelinguistic children
     cannot. However, more recent work seems to have made the significance
     of these experiments less clear. For the case of humans, see Kristin
     Ratliff and Nora Newcombe, "Is Language Necessary for Human Spa-
     tial Reorientation? Reconsidering Evidence from Dual Task Paradigms,"
     Cognitive Psychology 56 (2008): 142–63. Giorgio Vallortigara has also re-
     ported that chickens can solve the task that gave rats so much trouble; see
     Vallortigara et al., "Reorientation by Geometric and Landmark Infor-

mation in Environments of Different Size," *Developmental Science* 8 (2005): 393–401.

144 *But one plausible model, drawing on the work of several people*: Daniel Dennett's *Consciousness Explained* (New York: Little, Brown and Co., 1991) is an important source for the outlines of the view. For the idea that inner speech originates in repurposed efference copies, see Simon Jones and Charles Fernyhough, "Thought as Action: Inner Speech, Self-Monitoring, and Auditory Verbal Hallucinations," *Consciousness and Cognition* 16, no. 2 (2007): 391–99. Peter Carruthers suggests that inner speech is a means for internal "broadcast" that facilitates deliberate, rational styles of thought in his paper "An Architecture for Dual Reasoning," in Jonathan Evans and Keith Frankish, eds., *In Two Minds: Dual Processes and Beyond* (Oxford and New York: Oxford University Press, 2009). Fernyhough's book about inner speech, *The Voices Within*, will appear from Basic Books in 2016. My thinking about inner speech has also been influenced by Kritika Yegnashankaran's PhD thesis, "Reasoning as Action," Harvard University, 2010.

144 *I'll now tie these familiar facts to a concept that has become increasingly important*: I'll say more about the framework that introduced this concept shortly. Good sources are the Merker paper cited earlier: "The Liabilities of Mobility: A Selection Pressure for the Transition to Consciousness in Animal Evolution," *Consciousness and Cognition* 14 (2005): 89–114; and Kalina Christoff et al., "Specifying the Self for Cognitive Neuroscience," *Trends in Cognitive Sciences* 15, no. 3 (2011): 104–12.

144 *Without using the term, I introduced the idea of efference copies*: I also discussed one of the phenomena that efference copies are (probably) important in explaining: perceptual constancies. For example, when our eyes jump about (as they routinely do), objects appear to remain stable. This is one aspect of the family of "constancy" phenomena; other aspects include our ability to compensate for changes in lighting conditions, something that does not involve actions and efference copies. The role played by efference copies in constancy phenomena is still being worked out. See W. Pieter Medendorp, "Spatial Constancy Mechanisms in Motor Control," *Philosophical Transactions of the Royal Society B* 366 (2011): 20100089.

146 *Within a terminology used by Daniel Kahneman and other psychologists, it's a means for System 2 thinking*: Kahneman's book *Thinking, Fast and Slow* (New York: Farrar, Straus and Giroux, 2011) is already a classic. See also Evans and Frankish's edited collection of papers, *In Two Minds: Dual Processes and Beyond*. Dewey did place much emphasis on the imagined rehearsal of actions, especially in his theory of moral behavior.

147 *Gesturing to the careening inner monologues*: See his *Consciousness Explained*. Dennett does not make use of efference copies in his model. He ties his account of the origin of the Joycean machine to Richard Dawkins's notion of the transmission of *memes*, an idea about which I'm more skeptical (see Dawkins, *The Selfish Gene*, Oxford and New York: Oxford University Press, 1976).

147 *In an experiment done in 2001*: See Harald Merckelbach and Vincent van de Ven, "Another White Christmas: Fantasy Proneness and Reports of 'Hallucinatory Experiences' in Undergraduate Students," *Journal of Behavior Therapy and Experimental Psychiatry* 32, no. 3 (2001): 137–44.

148 *In landmark work from the 1970s, the British psychologists Alan Baddeley*: See Alan Baddeley and Graham Hitch, "Working Memory," in *The Psychology of Learning and Motivation*, Vol. VIII, ed. Gordon H. Bower, 47–89 (Cambridge, MA: Academic Press, 1974).

150 *A second-generation version of the workspace theory*: See Stanislas Dehaene and Lionel Naccache, "Towards a Cognitive Neuroscience of Consciousness: Basic Evidence and a Workspace Framework," *Cognition* 79 (2001): 1–37.

151 *A phenomenon that has for a long time appeared to have some connection to consciousness*: See especially the work of David Rosenthal, such as "Thinking That One Thinks," in Martin Davies and Glyn Humphreys, eds., *Consciousness: Psychological and Philosophical Essays*, 197–223 (Oxford: Blackwell Publishing, 1993).

152 *No one knows how old human language is*: See W. Tecumseh Fitch, *The Evolution of Language* (Cambridge, U.K.: Cambridge University Press, 2010).

153 *In 1950, the German physiologists Erich von Holst and Horst Mittelstaedt*: See Von Holst and Mittelstaedt, "The Reafference Principle (Interaction Between the Central Nervous System and the Periphery)," 1950, reprinted in *The Behavioural Physiology of Animals and Man*: *The Collected Papers of Erich von Holst*, vol. 1, trans. Robert Martin, 139–73 (Coral Gables, FL: University of Miami Press, 1973).

In one respect, the terminology I take from them is not the best. The internal signals used to deal with reafference need not be *copies*, in any normal sense, of the output signal sent to the muscles. What I call *efference copies* are sometimes called *corollary discharges* instead. The term "discharge" is more neutral than "copy." Trinity Crapse and Marc Sommer, in "Corollary Discharge Across the Animal Kingdom," *Nature Reviews Neuroscience* 9 (2008): 587–600, argue that efference copies should be seen as one *kind* of corollary discharge. That's perhaps a good way of setting things up. However, in this book I want to take advantage of the whole network of distinctions that von Holst and Mittelstaedt introduced: *afference* versus *efference, reafference* versus *exafference,* and so on. The word "copy" has become standard in this framework, so I stay with it.

These phenomena were first studied in the case of vision, and versions of the main idea—the need to compensate for reafference in order to resolve ambiguity in perception—were introduced in theories of vision that date back to the seventeenth century. An interesting historical sketch is in Otto-Joachim Grüsser, "Early Concepts on Efference Copy and Reafference," *Behavioral and Brain Sciences* 17, no. 2 (1994): 262–65.

155 *But memory of this kind* is *a communicative phenomenon*: I discuss this in "Sender-Receiver Systems Within and Between Organisms," *Philosophy of Science* 81 (2014): 866–78.

7. Experience Compressed

161 *Why don't we* all *live for a longer time?*: The classic works on aging drawn on in this chapter are Peter Medawar, *An Unsolved Problem of Biology* (London: H. K. Lewis and Company, 1952); George Williams, "Pleiotropy, Natural Selection, and the Evolution of Senescence," *Evolution* 11, no. 4 (1957): 398–411; and William Hamilton, "The Moulding of Senescence by Natural Selection," *Journal of Theoretical Biology* 12, no. 1 (1966): 12–45. A good review of the development of the evolutionary theory of aging is Michael Rose et al., "Evolution of Ageing since Darwin," *Journal of Genetics* 87 (2008): 363–71. A theory of aging I don't discuss explicitly is the *disposable soma* theory. I see this as a variant on the Williams theory. This is discussed by Thomas Kirkwood in "Understanding the Odd Science of Aging," *Cell* 120, no. 4 (2005): 437–47, another good review of all these issues.

169 *Hamilton died in 2000, after catching malaria*: The quote is from "My Intended Burial and Why," *Ethology Ecology and Evolution* 12, no. 2 (2000): 111–22. For more by this remarkable thinker, see *Narrow Roads of Gene Land: The Collected Papers of W. D. Hamilton,* Volume 1: *Evolution of Social Behaviour* (Oxford and New York: W. H. Freeman/Spektrum, 1996). In the end he was buried near Oxford, with an inscription from his partner on a bench nearby, noting that in time, carried by a drop of rain, he'll make it to the Amazon.

170 *The evolutionary theory of aging gives us an explanation for the basic facts*: This theory does not specify *how* age-related breakdown will take

place, though as Williams noted, it predicts that many different problems will arise as one gets older. Biologists still explore general mechanisms by which the decline occurs—either in mammals or in a wider range of organisms. Some hypotheses that posit a single widespread source of breakdown may be partial rivals to the evolutionary theory of aging as described here. It's sometimes hard to tell which theories are rivals and which are compatible with each other. For a recent study of mechanisms, see Darren Baker et al., "Naturally Occurring p16^{Ink4a}-Positive Cells Shorten Healthy Lifespan," *Nature* 530 (2016): 184–89.

170 *Female octopuses, in general, are an extreme case of semelparity*: See Jennifer Mather, "Behaviour Development: A Cephalopod Perspective," *International Journal of Comparative Psychology* 19, no. 1 (2006): 98–115.

171 *There's at least one exception*: See Roy Caldwell, Richard Ross, Arcadio Rodaniche, and Christine Huffard, "Behavior and Body Patterns of the Larger Pacific Striped Octopus," *PLoS One* 10, no. 8 (2015): e0134152. The paper does not describe this octopus as "iteroparous," as earlier studies did: "LPSO [their octopus] appear better designated as 'continuous spawning' with a single prolonged egg-laying period, rather than 'iteroparous' with multiple discretely separate egg-laying periods."

173 *Then the shells were abandoned*: Again see Kröger, Vinther, and Fuchs, "Cephalopod Origin and Evolution: A Congruent Picture Emerging from Fossils, Development and Molecules," *BioEssays* 33 (2011): 602–13.

174 *In 2007 they were inspecting a rocky outcrop*: See Bruce Robison, Brad Seibel, and Jeffrey Drazen, "Deep-Sea Octopus (*Graneledone boreopacifica*) Conducts the Longest-Known Egg-Brooding Period of Any Animal," *PLoS One* 9, no. 7 (2014): e103437.

175 *As a result, evolution has tuned its lifespan differently*: Another likely exception to the short-lived cephalopod rule is the vampire squid. Despite the name, this is not a very frightening animal. So little is known about these creatures' lives that a Dutch scientist, Henk-Jan Hoving, and some collaborators recently started to study old laboratory specimens, preserved for years in dusty bottles, to get some clues. They found evidence that unlike nearly all other cephalopods, female vampire squid go through multiple reproductive cycles, with considerable spacing between them. They think the cycle is likely to be repeated more than twenty times. If this is right, they must have long lives. They, too, are deep-sea animals, living in the cold and the metabolic slowdown of the depths. We don't have any evidence bearing directly on the predation risks they face. See Henk-Jan Hoving, Vladimir Laptikhovsky, and

Bruce Robison, "Vampire Squid Reproductive Strategy Is Unique among Coleoid Cephalopods," *Current Biology* 25, no. 8 (2015): R322–23.

175 *Putting these things together*: In one respect my treatment of cephalopod aging in this chapter is quite unorthodox. I am applying mainstream theoretical ideas (Medawar, Williams, etc.), but octopuses have for some time been seen as a problem for these very ideas. This is because it has appeared to many people that octopuses are "programmed" to die at a certain stage. Their breakdown has seemed orderly and "planned"—all these terms are frequently used about octopus death. When lists of cases are given that might be problems for the Medawar-Williams theory, octopuses are usually prominent on the list. The Medawar-Williams view does *not* see age-related breakdown as happening "by design," but octopuses do give this impression.

Buttressing this view is a 1977 study of the physiological basis for octopus senescence: Jerome Wodinsky's "Hormonal Inhibition of Feeding and Death in Octopus: Control by Optic Gland Secretion," *Science* 198 (1977): 948–51. The paper reports that death in the species *Octopus hummelincki* is caused by some secretion(s) from the "optic glands." When these glands are removed, octopuses of both sexes live longer and behave differently. In Wodinsky's interpretation: "The octopus apparently possesses a specific 'self-destruct' system." Why might they have such a thing? Wodinsky offers a hypothesis in a footnote: "in both sexes, this mechanism guarantees the elimination of old, large predatory individuals and constitutes a very effective means of population control."

If this claim about population control is intended as an explanation of *why* this death-causing mechanism exists, then it's apparently in conflict with the general principles about evolution I draw on early in this chapter. Suppose a mutant arose that lived for longer and gained a few extra matings. The fact that it might harm the population will not prevent that mutant from becoming more common. It is very hard for "population control" measures not to be subverted by free riders.

A modeling paper by Justin Werfel, Donald Ingber, and Yaneer Bar-Yam argues that it *is* possible for programmed death, of the sort often associated with octopuses, to evolve. The paper is called "Programmed Death Is Favored by Natural Selection in Spatial Systems," *Physical Review Letters* 114 (2015): 238103. The model used in this paper is one in which reproduction and dispersal is a local affair, though: a parent's offspring tend to settle and grow up nearby. That can cause problems of competition within a family (your offspring and perhaps grand-offspring are competing among themselves for the same local resources).

A range of models since the 1980s have shown that this sort of situation, where "an apple doesn't fall far from the tree," can have special evolutionary consequences. However, octopuses don't reproduce in that way. When an egg hatches, the larva joins the plankton and drifts away, and then settles on the sea floor somewhere, if it survives. Benjamin Kerr and I put together a model of cooperative behaviors in cases like this: Godfrey-Smith and Kerr, "Selection in Ephemeral Networks," *American Naturalist* 174, no. 6 (2009): 906–11. As far as anyone knows, young octopuses do not have a way to settle near where their mothers lived. If they did (by some sort of chemical tracking), that would have a number of interesting consequences, including the possibility of cooperation and reproductive "restraint."

I think octopus death is probably less "programmed" than it looks, though, and is some sort of extreme manifestation of the phenomena recognized by the Medawar-Williams theory. (See the Kirkwood paper I cited for further arguments of this kind, though not aimed at the octopus case.) The Wodinsky paper contains some clues. Removal of the optic glands causes a range of behavioral changes, as well as delayed senescence ("When these glands are removed after eggs have been laid, the female ceases to brood the eggs, begins to eat again, gains weight, and lives for a prolonged period"). The glands, when present, may cause not senesence per se, but a behavioral and physiological profile that has senesence as a by-product.

In one way, cephalopods are good cases for the evolutionary theory of aging: their predation risk is acute, and their lives are so short. In another way, they seem like bad cases: their breakdown looks too orderly, too "programmed." Perhaps there is something missing from the story I've told here—especially in male octopuses, who do not brood eggs, the sudden decline looks odd. But "population regulation" is unlikely, and I think the Medawar-Williams-Hamilton theory will turn out to apply.

8. Octopolis

179 *These days, the main place I watch octopuses*: An initial summary of the site's unusual features is in Godfrey-Smith and Lawrence, "Long-Term High-Density Occupation of a Site by *Octopus tetricus* and Possible Site Modification Due to Foraging Behavior," *Marine and Freshwater Behaviour and Physiology* 45 (2012): 1–8. The site continues to change. Updates can be found on the website Metozoan.net.

179 *Reports of clumps of octopuses had cropped up*: In one of our papers, we include a table that categorizes previous reports of clumps and social interaction in octopuses. See Table 1 in Scheel, Godfrey-Smith, and Lawrence, "Signal Use by Octopuses in Agonistic Interactions," *Current Biology* 26, no. 3 (2016): 377–82.

180 *As far as we can tell, they behave fairly similarly*: We can't be entirely sure about this, because the cameras themselves are a temporary addition to their environment. The cameras sit on tripods, often quite close to the animals. Sometimes a camera is attacked by an octopus. Our impression is that much of the time, the activities caught on camera with divers absent are not very different from what goes on when divers are down with them, and that most of the time the cameras are not the focus of much octopus attention. But it's hard to be sure.

181 *David did his training studying lions in Africa*: See, for example, Scheel and Packer, "Group Hunting Behavior of Lions: A Search for Cooperation," *Animal Behaviour* 41, no. 4 (1991): 697–709.

184 *Sometimes one of our unmanned cameras will film an octopus*: I can't be sure about these cases, as there might be an octopus out of sight (behind the camera). Or the camera itself might prompt some of these behaviors.

184 *Sometimes he will raise his mantle*: The object in the background is one of our video cameras on a tripod. This tripod is of a tall kind we started using recently in one position. The other tripods are low and less conspicuous.

185 *He noticed that the darkness of skin color is a reliable predictor*: See Scheel, Godfrey-Smith, and Lawrence, "Signal Use by Octopuses in Agonistic Interactions."

185 *I commissioned an artist*: The drawing was done by Eliza Jewett. A version of this image also appeared in Scheel, Godfrey-Smith, and Lawrence, "Signal Use by Octopuses in Agonistic Interactions."

186 *In 1982 Martin Moynihan and Arcadio Rodaniche reported*: This is the same paper discussed in chapter 5: "The Behavior and Natural History of the Caribbean Reef Squid (*Sepioteuthis sepioidea*). With a Consideration of Social, Signal and Defensive Patterns for Difficult and Dangerous Environments," *Advances in Ethology* 25 (1982): 1–151.

187 *The paper by Caldwell, Ross, and colleagues*: See Caldwell et al., "Behavior and Body Patterns of the Larger Pacific Striped Octopus," *PLoS One* 10 (2015): e0134152.

190 *Our second paper about the site*: This is Scheel, Godfrey-Smith, and Lawrence, "*Octopus tetricus* (Mollusca: Cephalopoda) as an Ecosystem Engineer," *Scientia Marina* 78, no. 4 (2014): 521–28.

191 *In 2011, a study of a species closely related to our Octopolitans*: See Elena Tricarico et al., "I Know My Neighbour: Individual Recognition in *Octopus vulgaris*," *PLoS One* 6, no. 4 (2011): e18710.

191 *A more controversial study from 1992*: This is Graziano Fiorito and Pietro Scotto, "Observational Learning in *Octopus vulgaris*," *Science* 256 (1992): 545–47.

194 *The common ancestor of birds and humans*: See Dawkins, *The Ancestor's Tale* (New York: Houghton Mifflin, 2004).

195 *In a famous paper from 1972, Andrew Packard argued*: See his "Cephalopods and Fish: The Limits of Convergence," *Biological Reviews* 47, no. 2 (1972): 241–307. See also Frank Grasso and Jennifer Basil, "The Evolution of Flexible Behavioral Repertoires in Cephalopod Molluscs," *Brain, Behavior and Evolution* 74, no. 3 (2009): 231–45.

195 *The new view has it that the most recent common ancestor*: Here again I use Kröger, Vinther, and Fuchs, "Cephalopod Origin and Evolution: A Congruent Picture Emerging from Fossils, Development and Molecules," *Bioessays* 33 (2011): 602–13. There is some uncertainty about where the *Vampyromorpha*, the "vampire squid," fit in. Note also that "decapod" can refer to a group of crustaceans, as well as a group of cephalopods.

196 *There still might have been competition*: Not only has the dating of cephalopod origins changed since Packard's time; the same has happened with fish. The sorts of fish he saw as the cephalopod competitors are now thought to have evolved earlier than he thought, perhaps in the Permian, which is around the new date for the common ancestor of coleoid cephalopods. See Thomas Near et al., "Resolution of Ray-Finned Fish Phylogeny and Timing of Diversification," *Proceedings of the National Academy of Sciences* 109, no.34 (2012): 13698–703.

196 *In 2015 the first octopus genome was sequenced*: See Caroline Albertin et al., "The Octopus Genome and the Evolution of Cephalopod Neural and Morphological Novelties," *Nature* 524 (2015): 220–24.

197 *An example is a recent study of memory*: See Christelle Jozet-Alves, Marion Bertin, and Nicola Clayton, "Evidence of Episodic-like Memory in Cuttlefish," *Current Biology* 23, no. 23 (2013): R1033–35. The bird study they modeled their work on is one cited above: Clayton and Dickinson, "Episodic-like Memory During Cache Recovery by Scrub Jays," *Nature* 395 (2001): 272–74.

201 *But in 2002 one small bay was designated a marine sanctuary*: The sanctuary is at Cabbage Tree Bay, in the north of Sydney.

201 *By the middle 1800s the fishers in the North Sea began to wonder*: Here I use especially Charles Clover's book *The End of the Line: How Over-*

fishing Is Changing the World and What We Eat (New York: New Press, 2006). Alanna Mitchell's *Sea Sick: The Global Ocean in Crisis* (Toronto: McClelland and Stewart, 2009) is similarly alarming reading. A shorter piece that's very good (again alarming) is Elizabeth Kolbert's "The Scales Fall," *The New Yorker*, August 2, 2010. Huxley's speech was given at the Fisheries Exhibition in London in 1883. Clover says, "Another parliamentary inquiry of which the ailing Huxley was a member reversed those conclusions within the decade."

201 *Within a few decades many of these fisheries*: In the case of cod, the decline had apparently already started by the time of Huxley's 1883 speech. It accelerated but then halted during World War I. After the war, cod stocks fluctuated, but declined, and then in 1992 the fishery collapsed completely on the Canadian side. Data from 2015 suggest that the fish is now doing much better, thanks to reduced fishing ("Cod Make a Comeback . . . ," *New Scientist*, July 8, 2015).

202 *One example is acidification*: I've not found a lot of work about cephalopods and ocean acidification. Some fairly worrying data are discussed in H. O. Pörtner et al., "Effects of Ocean Acidification on Nektonic Organisms," which appears in *Ocean Acidification*, edited by J.-P. Gattuso and L. Hansson (Oxford: Oxford University Press, 2011). Roger Hanlon, quoted by Katherine Harmon Courage, says that although cephalopods can handle various kinds of "dirty" water, they are very sensitive to the level of acidity (pH), because of their peculiar blood chemistry, and acidification is a serious threat to them. See Katherine Harmon Courage, *Octopus! The Most Mysterious Creature in the Sea* (New York: Current/Penguin, 2013), 70 and 213.

203 *When I asked Barron*: For a write-up of some of these ideas, see Andrew Barron, "Death of the Bee Hive: Understanding the Failure of an Insect Society," *Current Opinion in Insect Science* 10 (2015): 45–50.

203 *In many parts of the world's seas there are "dead zones"*: See Alanna Mitchell's *Sea Sick: The Global Ocean in Crisis*, and, for a summary, "What Causes Ocean 'Dead Zones'?," *Scientific American*, September 25, 2102, www.scientificamerican.com/article/ocean-dead-zones. According to Mitchell's book, the number of such zones has doubled every decade since 1960.

ACKNOWLEDGMENTS

I am grateful to the many scientists—marine biologists, evolutionary theorists, neuroscientists, and paleontologists—who helped me with this project. First on the list must be Crissy Huffard and Karina Hall, who aided and encouraged my earliest efforts to understand cephalopods. Other biologists who made important contributions include Jim Gehling, Gáspár Jékely, Alexandra Schnell, Michael Kuba, Jean Alupay, Roger Hanlon, Jean Boal, Benny Hochner, Jennifer Mather, Andrew Barron, Shelley Adamo, Jean McKinnon, David Edelman, Jennifer Basil, Frank Grasso, Graham Budd, Roy Caldwell, Susan Carey, Nicholas Strausfeld, and Roger Buick. The role of my companions at Octopolis—Matt Lawrence, David Scheel, and Stefan Linquist—will be clear from the text. I am grateful to them also for permission to use several images from videos we obtained together.

On the philosophical side, the influence of Daniel Dennett's work will be evident to the many admirers of his books, and I am grateful also to Fred Keijzer, Kim Sterelny, Derek Skillings, Austin Booth, Laura Franklin-Hall, Ron Planer, Rosa Cao, Colin Klein, Robert Lurz, Fiona Schick, Michael Trestman, and Joe Vitti. Dive Center Manly and Let's Go Adventures (Nelson Bay) provided invaluable support to the scuba diving. Eliza Jewett drew the figures on pages 49 and 186, and Ainsley Seago drew the figure on page 46. The photograph on the first page of the color insert appears also in "Cephalopod Cognition (book review)," *Animal Behaviour*, vol. 106, August 2015, pp. 145–47. Further thanks are due to Denise Whatley, Tony Bramley, Cynthia Chris, Denice Rodaniche, Mick Saliwon, and Lyn Cleary. The City University of New York Graduate Center is a marvelous place to do academic work, a place that gives freedom to think and write, along with a superb intellectual atmosphere. Deep thanks are due to all the custodians of Cabbage Tree Bay Aquatic Reserve, Booderee National Park, Jervis Bay Marine Park, and Port Stephens–Great Lakes Marine Park for their work in protecting and caring for these ecosystems.

Alex Star's role in this project was pivotal in ways that reach far beyond the familiar indispensability of a good editor. Finally, I must thank Jane Sheldon, who made acute comments on several drafts, spotted notable animals in the sea, inspired and contributed many ideas, and dealt patiently with the steadily increasing quantities of salt water and neoprene that permeate our tiny apartment on the coast where this book was conceived.

INDEX

Page numbers in *italics* refer to illustrations.